COUNTRIES

of

INVENTION

Countries of Invention

© 1993, Addison-Wesley Publishers Ltd. and Rubicon Publishing Inc.

ISBN 0-201-60218-0

FOR JOHN AND GAYLE SMALLBRIDGE

Canadian Cataloguing in Publication Data

Main entry under title:

Countries of invention : world contemporary
    literature

Includes bibliographical references

ISBN 0-201-60218-0

1. Literature - Collections   I. Stephenson, Craig

PN6014.C68 1993   808.8'004   C92-094781-6

Designer: *Wycliffe Smith*

Editors: *Mei-lin Cheung, Elizabeth Siegel Masih*

Editorial Assistant: *Shen-Yi Goh*

The Publishers gratefully acknowledge the following reviewers for their insights
and advice.

Bil Chinn, Edmonton Public Schools, Alberta

Irene Kent, Peel Board of Education

Bryant Knox, Vancouver Board of Education

Tom Rossiter, Roman Catholic School Board of St. John's, Nfld.

Ken Roy, Etobicoke Board of Education, Ontario

Jone Schoeffel, Appleby College, Ontario

COVER ART: "The Castle in the Pyrenees," 1961, by Rene Magritte.
© 1992 C. Herscovici/ARS, New York

93 94 95 96   5 4 3 2 1

COUNTRIES

of

INVENTION

Edited by
CRAIG STEPHENSON

With an Introduction by
Alberto Manguel

Addison-Wesley Publishers Limited

Rubicon Publishing Inc.

TABLE OF

CONTENTS

PREFACE

Timothy Findley's essay "The Countries of Invention" (which appeared in the Spring 1984 issue of *Canadian Literature*) provides the title for this anthology of contemporary writing. The essay serves as a touchstone and unites diversity of international writings between the covers of this volume in three important ways.

First, stories admonish us in much the same way as Findley's mentor, the American writer Thornton Wilder, would admonish him: "There's nothing down there but your feet, Findley. Look around you; it's more interesting." The writers I've selected for this collection share this urgent, far-sighted attentiveness. This is why Ben Okri compares writing to practising a martial art: "you pay for every mistake you make ... for loss of attention, ... for taking anything for granted." And this is why, having invented an intensely disturbing seascape of light and shadow, and movement, Annie Dillard asks: "Have we rowed out to the thick darkness, or are we all playing pinochle in the bottom of the boat?" These writers invite us as readers to join them in the boat and then row out toward the blackness, all the while speaking with Wilder's urgent tone that says: "Pay attention."

Second, Wilder also provides "a perfect example" of the kind of imaginative freedom which no writer should be asked to relinquish: the diverse, imagined worlds which he created resonate with life. Caesar's Rome in *The Ides of March*, eighteenth century Peru in *The Bridge of San Luis Rey*, Noah's Flood in *The Skin of Our Teeth*, and beyond the grave in *Our Town* "feel more real than any place we see and hear in our everyday lives," more authentic. We now live in a time when readers are being asked to judge the authenticity of a writer's voice by considering the writer's race or gender as much as the text. I acknowledge the importance of these factors, but instead I have preferred to use the example of Wilder's "authenticity" and imaginative freedom as I have collected these stories. The writers whose works have been chosen here are a multifarious international group of women and men, but it would be a mistake to identify their countries of invention with their countries of origin, or to cull from their imaginations a special cultural or gender perspective. In fact, they themselves will often confound any such attempts: Gabriel Garcia Marquez records the inner monologue of a young girl watching it rain; Yukio Mishima views a Japanese sea through the eyes of an elderly French pilgrim; Tatyana Tolstaya invents the buckling emotional life of a young man whom we might call "mentally handicapped;" and, retelling an ancient Taoist tale, Marguerite Yourcenar imitates an old Chinese style of brushwork and yet deftly paints her own landscape.

Finally, we should consider that the writer is a visitor who goes to the countries of invention in search of questions, not answers. Consequently, I have trusted most those writers whose works engender questions. But be forewarned: this means the answers lie elsewhere. It is the reader who writes the story: the answers to the questions can be found within each of us as we read through to the last word. A kind of legacy of invention is hereby passed on.

Thanks, then, to Timothy Findley for the gift of the title; and a special thanks to Alberto Manguel for his generosity and support.

Craig Stephenson

INTRODUCTION

ALBERTO MANGUEL

For reasons that are still mysterious and which, perhaps, if revealed, might seem banal, in the eighth year A.D. the poet Publius Ovidius Naso, one of the greatest writers of antiquity, was banished from Rome by the Emperor Augustus. Ovid (his three names reduced to one by centuries of devoted readers) ended his days in a backwater village on the west coast of the Black Sea, pining for Rome. He had been at the heart of the empire which, in those days, was synonymous with the world; to be banished was for Ovid like a death sentence, because he could not conceive of life outside his beloved city. According to Ovid himself, at the root of the imperial punishment was a poem. We don't know what the words of that poem were, but they were powerful enough to terrify an emperor.

Since the beginning of time (the telling of which is also a story) we have known that words are dangerous creatures. In Babylon, in Egypt, in ancient Greece, the person capable of inventing and recording words, the writer, whom the Anglo-Saxons called "the maker," was thought to be the darling of the gods, a chosen one on whom the gift of writing had been bestowed. According to Socrates, in a legend which he either retold or imagined, the art of writing was the creation of the Egyptian god Theuth, who also invented mathematics, astronomy, checkers and dice. In offering his invention to the Pharaoh, Theuth explained that his discovery provided a recipe for memory and wisdom. But the Pharaoh wasn't convinced: "What you have discovered," he said, "is not a recipe for memory but for reminder. And it is not true wisdom that you offer your disciples, but only its semblance, for by telling them of many things without teaching them, you will make them seem to know much, while for the most part they know nothing, and are filled not with wisdom but with the conceit of wisdom." Ever since, writers and readers have debated whether writing effectively achieves anything. Some, agreeing with Theuth, believe that we can learn from writing, that it makes us wise by granting us the memory of centuries of experience, a memory that allows us to survive in times of adversity. Others, agreeing with the Pharaoh, say with the poet W.H. Auden that "poetry makes nothing happen;" that the memory preserved in writing does not inspire wisdom, that we learnt nothing through the imagined word and that times of adversity are proof of writing's failure.

It is true that, confronted with the blind imbecility with which we have tried to destroy our planet, the relentlessness with which we inflict pain on ourselves and others, the extent of our greed and cowardice and envy, the arrogance with which we strut among our fellow living creatures, it is hard to believe that writing — literature or any other art, for that matter — teaches us anything. If after reading lines such as Walt Whitman's,

> I celebrate myself, and sing myself
> And what I assume you shall assume,
> For every atom belonging to me as good as belongs to you

we are still capable of all such atrocities, then perhaps the writer does make nothing happen.

In at least one sense, all writing is memory: all literature preserves something which otherwise would die away with the flesh and bones of the writer. This memory of writing is infinite. Humans can remember little: even extraordinary memories such as that of Cyrus, king of the Persians, who could call every soldier in his armies by name, are nothing compared to the volumes that fill our seemingly endless libraries. Whether on stone, wax tablets, knotted cords of the Incas, wampum belts in North America, paper, or computer disks, the writer sets down for us a certain vision of reality, and the sum total of these visions is almost as vast as the universe it attempts to depict. Among these visions are accounts of our atrocities, and in that sense all literature is testimonial. But among the testimonies are also reflections on those atrocities so that they are not allowed to take place in silence. They are reminders of better things, of hope and consolation and compassion, and hold the implication that of these too, we are all of us capable. Not all of these we achieve, and none of these we achieve all the time. But literature reminds us that they are there, these human qualities, following our horrors as certainly as birth succeeds death. They too define us.

The writers whose work Craig Stephenson has collected in this anthology may seem, at first glance, curiously diverse. What they have to say colours, of course, their various styles, but the language in which they say it influences in turn what they say. Language is never neutral: political traditions and power games have a hand in the construction of language, just as much as music and logic, and no writer is ever innocent in the use of words. The apparent diversity of these texts is therefore due both to the choices the writer makes and to the language to which, so to speak, the writer belongs. Translation (that is to say, the art of recreating in another language the essence of a foreign text) tries to overcome these particularities — and on occasion succeeds. However, in spite of these diversities, the writers in *Countries of Invention* share a common characteristic: all, without exception, create out of words landscapes, exterior and interior, through which we, the readers, can wander. They are all different landscapes, shaped as I have said by the features of a specific subject and the strength of a specific language, but they are all landscapes which readers must explore at their own risk and peril. Some readers may find themselves best suited to the lush landscape of a Garcia Marquez or a Marguerite Yourcenar; others may feel more akin to the sober kitchens and bedrooms of an Alice Munro. Some will prefer the zoological realms of an Annie Dillard; others will drift towards the bleak provinces of a Tatyana Tolstaya or a Cynthia Ozick. Some readers, the luckiest ones, will be citizens of the world. And what every reader will find is that the common trait of these writers is that they are all, to a greater or lesser extent, magicians.

Words, the squiggles that dance before our eyes on the page and the sounds they create in the darkness of the mind, are essentially a form of magic. From literally thin air, the words the writer puts down allow us to discover, capture, explore, identify, transform, analyse, and ultimately inhabit the world around and within us. And sometimes almost understand it.

It is interesting that we, the readers, have come to recognize, over the centuries, the importance of literature in the mere act of living. And yet, paradoxically, little of that importance is admitted in our everyday life. One million viewers will follow the final episode of a sitcom on TV; fifty thousand spectators will cheer the Blue Jays during the opening game of the season; five hundred copies of a book by almost any of the best poets of our age will painfully be sold over one or two years. However, neither sitcom stars nor brilliant hitters are routinely exiled or imprisoned, tortured, or put to death for their work. Words, literature, books, because of their very nature relentlessly challenge the right of those in power, ask unsettling questions, put in doubt our assumptions. Literature may not be able to save anyone from injustice, but something about it must be perilously effective if every dictator, every totalitarian government, every threatened official tries to do away with it, by burning books, by banning books, by censoring books, by taxing books, by paying mere lip-service to the cause of literacy, by insinuating that reading is an elitist activity.

And it is curious that, in spite of thousands of years of experience, those in power haven't learned that their methods are ineffective. General Pinochet in Chile may have banned Don Quixote because he felt (quite rightly) that the book was a defence of an individual's civil liberties; Augustus in Rome might have exiled Ovid because he knew (and was probably not mistaken) that something in the poet's work accused him; every day, somewhere in the world, someone attempts (sometimes unsuccessfully) to stifle a book which plainly or obscurely sounds a warning. And again and again, empires fall and literature continues. Ultimately, the countries a writer invents — in the etymological sense of "to come upon," "to discover" — survive it all because they are the real world revealed under its true name. The rest, as we should have realized by now, is merely shadow without substance, the stuff of nightmares, and will vanish without a trace in the morning.

It is only natural: making my own collection, lifting my images from here and there — vistas, faces, gestures, accidents — carrying them forward with me, letting them rattle round in my brain, my innards until they have settled themselves, either as landmarks or as residents. I am a travelling country of invention. A roadshow.

> Timothy Findley,
> "The Countries of Invention"

"Some people," Miss R. said, "run to conceits or wisdom but I hold to the hard, brown, nutlike word. I might point out that there is enough aesthetic excitement here to satisfy anyone but a damned fool."

> Donald Barthelme,
> "The Indian Uprising"

What has LITERATURE got to do with it?

CHINUA ACHEBE

Chinua Achebe was born in 1930 in Ogidi, Eastern Nigeria. Having confronted the major social and political issues of contemporary Africa, he defines art as the effort to create a different reality from that which is given: "It is only the story that can continue beyond the war and the warrior.... It is only the story ... that saves our progeny from blundering like blind beggars into the spike of the cactus fence. The story is our escort; without it, we are blind. Does the blind own his escort? No, neither do we the story; rather it is the story that owns us."

MAN IS A GOAL-SETTING ANIMAL. ALONE OR IN concert with his fellows he does frequently tend to select and tackle his problems in graded priorities. He identifies personal goals, family goals, community goals, national and international goals; and he focuses his attention on solving them. At the national level, for example, he has invented short-term annual budgets and long-term five- or ten-year development plans; and, for good measure, we do have in this country chiliastic expectations such as health for all in the magic year of 2000.

Setting goals is a matter of intelligence and judgement. Faced with a confusing welter of problems all clamouring for solution at once, man's most rational strategy is to stay as cool as possible in the face of the confusion and attack the problems singly or in small manageable groups, one at a time. Of course the choice of what he must assault first or what he can reserve for last is of the utmost importance and can determine his success or failure.

The comprehensive goal of a developing nation like Nigeria is, of course, development, or its somewhat better variant, modernization. I don't see much room for argument about that. What can be, and is, vigorously debated is the quickest and safest route for the journey into modernization and what items should make up the traveller's rather limited baggage allowance.

But the problem with goals lies not only in the area of priorities and practicalities. There are appropriate and inappropriate goals, even wrong and unworthy goals. There are goals which place an intolerable strain on the pursuer. History tells us, for example, of leaders who in their obsessive pursuit of modernization placed on their people such pressures as they were unable to bear — Peter the Great of Russia, Muhammad Ali of Egypt, and others. Out of contemporary China rumours have come that the national goal of the one-child family which was set to combat a disastrous population problem has come into conflict with the desire of ordinary rural parents for male children and has apparently led to the large scale secret murder of female children. It is clear from these and similar examples that a nation might set itself a goal that puts its very soul at risk.

At the Tokyo Colloquium in October 1981, under the theme of Diversified Evolution of World Civilization, Professor Marion J. Levy of Princeton University, known for his study of the history of modernization in Japan and China, made the following remark about Japan:

> Well over half a century ago when everyone else was occupied with describing Japan in terms of the warrior and merchant classes, Yanagida Kunio took the position that the real heart of Japan was in the customs of the Japanese farmer.[1]

If Kunio was right the point made by Professor Levy is very instructive. The mercantile and militaristic (but particularly the militaristic) goals of Japan in the first half of this century would then seem to have been at variance with the real heart of Japan — or perhaps one should say that the heart of Japan was not fully in them. This is of course an area of discourse where firm proof and certainty would be unattainable. But I think that Kunio's view does gain credence from the fact that Japan, whose celebrated militarism suffered one of the most horrendous defeats ever visited on an army in modern, or indeed any, times, was yet able to survive and muster the morale to become in twenty-odd years a miracle of technological and economic success, outstripping all comers. A very colourful metaphor comes readily to mind — snatching victory out of the jaws of defeat.

The history of Nigeria from, say, 1970 to 1983 can be characterized by contrast as a snatching of defeat from the jaws of victory, if one considers how nearly 100 000 million *naira* went through our hands like so much sand through the fingers of a child at play on the beach. How do we begin to explain that? Did we not have goals? Did we not have development plans? Did we not have experts to guide our steps on the slippery slopes of modernization?

But we did have all those things — annual budgets, development plans, the lot. We were not short on experts, either. If we didn't have the particular kind we required, surely we had the money to hire him. What went wrong then? Our heart? Our mind? It seems our heart was not in it. Perhaps we suffered a failure of imagination. Perhaps psychologically we did not really wish to become a modern state; we saw the price of modernization and subconsciously decided we were not prepared to pay it.

Let us examine one or two of these suppositions, beginning with the question of the expert. No nation which contemplates modernization can neglect the role of the expert. He is needed; he must be paid for and he must be given adequate protection of tenure as well as respect so that others inferior to him may be motivated to strive and attain his expertise rather than hope through cheap politicking to manoeuvre themselves into his seat.

But having an expert among us does not absolve the rest of us from thinking. To begin with, the expert is generally an expert only in a narrow specialism. He can build a bridge for us perhaps, and tell us what weight of traffic it can support. But he can't stop us from hiring an attendant who will take a bribe and look the other way while the prescribed weight is exceeded. He can set up the finest machinery for us, but he can't create the technician who will stay at his post and watch the controls instead of going for a chat and some groundnuts under a mango tree outside.

So there is a limit to what an expert can do for us. In 1983, just before the overthrow of President Shagari's administration, I gave an interview for a television program which subsequently caused some offence in certain quarters. One of the questions put to me was what did I think about the President's Green Revolution program. And I said then, as I would say today, that it was a disaster which gave us plenty of food for thought and nothing at all in our stomach. Whereupon a certain fellow with a lot of grouse in him wrote in the papers that I should not have been asked to comment on agriculture because I was not an expert in that field. Well, we don't really need a Ph.D. in agriculture to tell us when our stomach is empty, do we? If we are in reasonable health we should all carry around with us reliable, inbuilt alarm systems popularly called hunger to apprise us of our condition!

I must say in this regard that the best experts do not themselves encourage us to have foolish and superstitious faith in their ability to solve our diverse development problems. In an essay published by the *American Economic Review* in 1984, Sir Arthur Lewis, one of the foremost development economists in the world and no stranger by any means to problems of African underdevelopment, did highlight in his inimitably elegant fashion the sheer plethora of prescriptions among development experts of differing persuasions:

> Every school has offered its own candidate for driver of the engine of growth. The physiocrats, agriculture; the Mercantilists, an export surplus; the classicists, the free market; the Marxists, capital; the neo-classicists, entrepreneurship; the Fabians, government; the Stalinists, industrialization; and the Chicago School, schooling.[2]

To sum up this marvellous passage I have composed a couplet which I beg pardon to inflict on you:

> There! we have it on the best authority
> Theorists of development cannot agree!

I will turn now to another world-famous economist, John Kenneth Galbraith, for a different kind of testimony. Interestingly, John Kenneth Galbraith is the current President of the American Academy of Arts and Letters. I must crave your indulgence to quote a fairly long extract from his address to the Academy in 1984 about the role of the arts in industry:

> Finally, let no one minimize the service that the arts render to established industry. In the years since World War II ... there has been no economic miracle quite like that of Italy. That lovely country has gone from one political disaster to another with one of the highest rates of economic growth of any of the industrial lands. The reason is not that the Italian

government is notably precise in its administration, the Italian engineers and scientists are better than others, that Italian management is inspired or that Italian trade unions are more docile than the AFL-CIO. The Italian success derives from the Italian artistic tradition. Italian products over the widest range are superior not in durability, not in engineering excellence, not in lower cost. They are better in design. Italian design and the consequent industrial success are the result of centuries of recognition of — including massive subsidy to — the arts.

In concluding his address, Galbraith made the following affirmation:

> The arts are not the poor relation of the economic world. On the contrary they are at the very source of its vitality.[3]

Before I leave these foreign references I must return very briefly to that other miracle, Japan, to which I have already made reference. If "there has been no economic miracle quite like that of Italy," there has been none to match Japan's in dramatic suddenness and awesomeness of scale. It has been uniquely salutary also for thoroughly debunking all the bogus mystique summoned to explain Western industrialism — the Protestant ethic, the Graeco-Judaic tradition, etc. We, the latecomers (as Marion J. Levy calls us), have every reason to pay special attention to Japan's success story as we take our faltering steps to modernization.

In the 1981 Tokyo Colloquium which I spoke about earlier we were attempting, among other things, to define the cultural ingredient, or as one of the Japanese scholars put it the "software," of modernization. One of the observations that made a particularly strong impact on me in this connection was a little family anecdote by Professor Kinichiro Toba of Waseda University:

> My grandfather graduated from the University of Tokyo at the beginning of the 1880s. His notebooks were full of English. My father graduated from the same university in 1920 and half of his notes were filled with English. When I graduated a generation later my notes were all in Japanese.

So ... it took three generations for us to consume Western civilization totally via the means of our own language.[4]

If Professor Toba's story is at all typical of the last 100 years of Japanese history (and we have no reason to believe otherwise), we can conclude that as Japan began the countdown to its spectacular technological lift-off it was also systematically recovering lost ground in its traditional mode of cultural expression. In one sense then it was travelling away from its old self towards a cosmopolitan, modern identity, while in another sense it was journeying back to regain a threatened past and selfhood. To comprehend the dimensions of this gigantic paradox and coax from it such unparalleled inventiveness requires not mere technical flair but the archaic energy, the perspective, the temperament of creation myths and symbolism.

It is in the very nature of creativity, in its prodigious complexity and richness, that it will accommodate paradoxes and ambiguities. But this, it seems, will always elude and pose a problem for the uncreative, literal mind (which I hasten to add is not the same as the literary mind, nor even the merely literate mind). The literal mind is the one-track mind, the simplistic mind, the mind that cannot comprehend that where one thing stands, another will stand beside it — the mind (finally and alas!) which appears to dominate our current thinking on Nigeria's need for technology.

The cry all around is for more science and less humanities (for in the narrow disposition of the literal mind more of one must mean less of the other). Our older universities have been pressured into a futile policy of attempting to allocate places on a 60:40 ratio in favour of science admissions. In addition, we have rushed to create universities of technology (and just as promptly proceeded to shut down half of them again) to demonstrate our priorities as well as confusions.

Nobody doubts that the modern developed world owes much of its success to scientific education and development. There is no doubt either that a nation can decide to emphasize science in its education program in order to achieve a specific national objective. When the Russians put the first man in orbit in 1961, John Kennedy responded by doubling United States space appropriations in 1962 and intensified a program of space research which was to land

Americans on the moon within the decade. But Kennedy did not ask the universities to starve out America's liberal arts education. As a matter of fact he had previously demonstrated sufficient awareness of the national need for the arts when at his inauguration he broke with tradition and gave pride of place to a reading by Robert Frost, the great New England poet.

Furthermore, it is important to realize that because a country like America with a well-developed and viable educational system may safely switch emphases around in its educational program it does not therefore follow that Nigeria, whose incipient program is already in a shambles, can do the same. What kind of science can a child learn in the absence, for example, of basic language competence and an attendant inability to handle concepts?

Have we reflected on the fact that in pre-independence Nigeria the only schools equipped adequately to teach science, namely the four or five government colleges, not only produced doctors and engineers like other schools but held an almost complete monopoly in producing novelists, poets, and playwrights?

Surely if this fact proves anything it is that education is a complex creative process and the more rounded it is the more productive it will become.[5] It is not a machine into which you feed raw materials at one end and pick up packaged products at the other. It is, indeed like creativity itself, "a many-splendoured thing."

The great nineteenth-century American poet Walt Whitman has left us a magnificent celebration of the many-sided nature of the creative spirit:

> Do I contradict myself?
> Very well then I contradict myself
> (I am large, I contain multitudes)...

The universal creative rondo revolves on people and stories. *People create stories create people*; or rather, *stories create people create stories*. Was it stories first and then people, or the other way round? Most creation myths would seem to suggest the antecedence of stories — a scenario in which the story was already unfolding in the cosmos before, and even as a result of which, man came into being. Take the remarkable Fulani creation story:

> In the beginning there was a huge drop of milk. Then the milk created stone; the stone created fire; the fire created water; the water created air.
>
> Then Doondari came and took the five elements and moulded them into man. But man was proud. Then Doondari created blindness and blindness defeated man ...

A fabulously rich story, it proceeds in stark successions of creation and defeat to man's death through hubris, and then to a final happy twist of redemption when death itself, having inherited man's arrogance, causes Doondari to descend a third time as Gueno the eternal one, to defeat death.

So important have such stories been to mankind that they are not restricted to accounts of initial creation but will be found following human societies as they recreate themselves through vicissitudes of their history, validating their social organizations, their political systems, their moral attitudes and religious beliefs, even their prejudices. Such stories serve the purpose of consolidating whatever gains a people or their leaders have made or imagine they have made in their existential journey through the world; but they also serve to sanction change when it can no longer be denied. At such critical moments new versions of old stories or entirely fresh ones tend to be brought into being to mediate the changes and sometimes to consecrate opportunistic defections into more honourable rites of passage.

One of the paradoxes of Igbo political systems is the absence of kings on the one hand, and on the other the presence in the language and folklore of a whole range of words for "king" and all the paraphernalia of royalty. In the Igbo town of Ogidi where I grew up I have found two explanation myths offered for the absence of kings. One account has it that once upon a time the title of king did exist in the community but that it gradually fell out of use because of the rigorous condition it placed on the aspirant, requiring him to settle the debt owed by every man and every woman in the kingdom.

The second account has it that there was indeed a king who held the people in such utter contempt that one day when he had a ritual kola-nut to break for them he cracked it between his teeth. So the people, who did not fancy eating kola-nut coated with the king's saliva, dethroned him and have remained republican ever since.

These are perhaps no more than fragmentary makeshift accounts though not entirely lacking in allegorical interest. There is for instance, a certain philosophical appropriateness to the point that a man who would be king over his fellows should in return be prepared personally to guarantee their solvency.

Be that as it may, those two interesting fragments of republican propaganda played their part in keeping kings' noses out of the affairs of Ogidi for as long as memory could go until the community, along with the rest of Nigeria, lost political initiative to the British at the inception of colonial rule. Thereafter a new dynasty of kings rose to power in Ogidi with the connivance of the British administration, thus rendering those mythical explanations of republicanism obsolete. Except perhaps that they may have left a salutary, moderating residue in the psyche of the new rulers and those they ruled.

I shall now, with your indulgence, present two brief parables from pre-colonial Nigeria which are short enough for the present purpose but also complex enough to warrant my classifying them as literature. I chose these two particularly because they stand at the opposite ends of the political spectrum.

Once upon a time, all the animals were summoned to a meeting. As they converged on the public square early in the morning one of them, the fowl, was spotted by his neighbours going in the opposite direction. They said to him, "How is it that you are going away from the public square? Did you not hear the town crier's summons last night?"

"I did hear it," said the fowl, "and I should certainly have gone to the meeting if a certain personal matter had not cropped up which I must attend to. I am truly sorry, but I hope you will make my sincere apologies to the meeting. Tell them that though absent in body I will be there with you in spirit in all your deliberations. Needless to say that whatever you decide will receive my whole-hearted support."

The question before the assembled animals was what to do in the face of a new threat posed by man's frequent slaughtering of animals to placate his gods. After a stormy but surprisingly brief debate it was decided to present to man one of their number as his regular sacrificial animal if he would leave the rest in peace. And it was agreed without a division that the fowl should be offered to man to mediate between him and his gods. And it has been so ever since.

The second story goes like this:

One day a snake was riding his horse coiled up, as was his fashion, in the saddle. As he came down the road he met the toad walking by the roadside.

"Excuse me, sir," said the toad, "but that's not the way to ride a horse."

"Really? Can you show me the right way, then?" asked the snake.

"With pleasure, if you will be good enough to step down a moment."

The snake slid down the side of his horse and the toad jumped with alacrity into the saddle, sat bolt upright and galloped most elegantly up and down the road. "That's how to ride a horse," he said at the end of his excellent demonstration.

"Very good," said the snake. "very good indeed; you may now come down."

The toad jumped down and the snake slid up the side of his horse and back into the saddle and coiled himself up as before. Then he said to the toad, "Knowing is good, but having is better. What good does fine horsemanship do to a fellow without a horse?" And then he rode away in his accustomed manner.[6]

On the face of it, those are just two charming animal stories to put a smile on the face or, if we are fortunate and have a generous audience, even a laugh in the throat. But beneath that admittedly important purpose of giving delight there lies a deep and very serious intent. Indeed, what we have before us are political and ideological statements of the utmost consequence revealing more about the societies that made and sustained them, and by which, in the reciprocal rondo of creativity, they were made and sustained, revealing far more than any number of political-science monographs could possibly ever tell us. We could literally spend hours analysing each story and discovering new significances all the time. Right now, however, we can take only a cursory look.

Consider the story of the delinquent fowl. Quite clearly it is a warning, a cautionary tale, about the danger to which citizens of small-scale democratic systems may be exposed when they neglect the cardinal duty of active participation in the political process. In such systems a man who neglects to lick his lips, as a certain proverb cautions us, will be asking the harmattan to lick them for him. It did for the fowl with a vengeance!

The second story is, if you will permit a rather predictable cliché, a horse of a different colour altogether. The snake is an aristocrat in a class society in which status and its symbols are not earned but ascribed. The toad is a commoner whose knowledge and expertise garnered through personal effort count for nothing beside the merit which belongs to the snake by some unspecified right such as birth or wealth. No amount of brightness or ability on the part of the toad is going to alter the position ordained for him. The few but potent words left with him by the snake embody a stern, utilitarian view of education which would tie the acquisition of skills to the availability of scope for their practice.

I have chosen those two little examples from Nigeria's vast and varied treasury of oral literature to show how such stories can combine in a most admirable manner the aesthetic qualities of a successful work of imagination with those homiletic virtues demanded of active definers and custodians of society's values.

But we must not see the role of literature only in terms of providing latent support for things as they are for it does also offer the kinetic energy necessary for social transition and change. If we tend to dwell more on stability it is only because society itself does aspire to, and indeed requires, longer periods of rest than of turmoil. But literature is also deeply concerned with change. That little fragment about the king who insulted his subjects by breaking their kola-nut in his mouth is a clear incitement to rebellion. But even more illuminating in this connection because of its subtlety is the story of the snake and the toad which at first sight may appear to uphold privileges but at another level of significance does in fact contain the seeds of revolution, the portents of the dissolution of an incompetent oligarchy. The brilliant makers of the story, by denying sympathetic attractivenesss to the snake, are exposing him in the fullness of time to the harsh tenets of a revolutionary justice.

I think I have now set a wide-ranging enough background to attempt an answer to the rhetorical question: What has literature got to do with it?

In the first place, what does "it" stand for? Is it something concrete like increasing the GNP or something metaphysical like the *It* which is the object of the quest in Gabriel Okara's novel, *The Voice?*

I should say that my "it" begins with concrete aspirations like economic growth, health for all, education which actually educates, etc., etc., but soon reveals an umbilical link with a metaphysical search for abiding values. In other words, I am saying that development or modernization is not merely, or even primarily, a question of having lots of money to spend or blueprints drawn up by the best experts available; it is in a critical sense a question of the mind and the will. And I am saying that the mind and the will belong first and foremost to the domain of stories. In the beginning was the Word, or the Mind, as an alternative rendering has it. It was the Word or the Mind that began the story of creation.

So it is with the creation of human societies. And what Nigeria is aiming to do is nothing less than the creation of a new place and a new people. And she needs must have the creative energy of stories to initiate and sustain that work.

Our ancestors created their different polities with myths embodying their varying perceptions of reality. Every people everywhere did the same. The Jews had their Old Testament on account of which early Islam honoured them as the people of the Book. The following passage appears in a brilliant essay in *Publications of the Modern Language Association of America*:

> The ideals that Homer portrayed in Achilles, Hector and Ulysses played a large role in the formation of the Greek character. Likewise when the Anglo-Saxons huddled around their hearth fires, stories of heroes like Beowulf helped define

them as a people, through articulating their values and defining their goals in relation to the cold, alien world around them.[7]

In the essay from which I took that passage the authors set out to demonstrate in detail the potentiality of literature to reform the self in a manner analogous to the processes of psychoanalysis: eliciting deep or unconsciously held primary values and then bringing conscious reflection or competing values to bear on them. The authors underscore the interesting point made by Roy Schafer that psychoanalysis itself is an essay into story-telling. People who go through psychoanalysis tell the analyst about themselves and others in the past and present. In making interpretations the competent analyst reorganizes and retells these stories in such a way that the problematic and incoherent self consciously told at the beginning of the analysis is sorted out to the benefit and sanity of the client. It would be impossible and indeed inappropriate to pursue this perceptive and tremendously important analogy between literature and psychoanalysis any further here, but I must quote its concluding sentence:

> ... if as Kohut, Meissner and others suggest the self has an inherent teleology for growth and cohesion, then literature can have an important and profound positive effect as well, functioning as a kind of bountiful, nourishing matrix for a healthy, developing psyche.[8]

This is putting into scientific language what our ancestors had known all along and reminds one of the common man who, on being told the meaning of "prose," exclaimed: "Look at that! So I have been speaking prose all my life without knowing it."

The matter is really quite simple. Literature, whether handed down by word of mouth or in print, gives us a second handle on reality; enabling us to encounter in the safe, manageable dimensions of make-believe the very same threats to integrity that may assail the psyche in real life; and at the same time providing through the self-discovery which it imparts a veritable weapon for coping with these threats whether they are found within problematic and incoherent selves or in the world around us. What better preparation can a people desire as they begin their journey into the strange, revolutionary world of modernization?

1 *Proceedings of the Tokyo Colloquium*, October 1981.

2 W. Arthur Lewis, "The State Development Theory," *American Economic Review*, 1984; reprinted in *Economic Impact*, 49, 1985, p. 82.

3 J.K. Galbraith, in *Proceedings: American Academy and Institute of Arts and Letters*, second series, no. 35, New York, 1984.

4 Proceedings of the Tokyo Colloquium, published in *The Daily Yomuiri*, 18 November 1981.

5 The Vice-Chancellor of Ibadan University, Professor Ayo Banjo, was reported as making the point that Ibadan does not teach mass communications and as yet "has produced most of the best writers in Nigerian journalism today": *Sunday Concord*, 16 February 1986.

6 Ulli Beier (ed.), *The Origin of Life and Death*, London, Heinemann Educational Books, 1966.

7 M.W. Alcorn and M. Bracher, "Literature, Psychoanalysis and the Reformation of the Self. A New Direction for Reader-Response Theory," *Publications of the Modern Language Association of America*, New York, May 1985, p. 350.

8 Ibid., p. 352

TWO WORDS

ISABEL ALLENDE

HER NAME WAS BELISA CREPUSCULARIO, NOT through baptism or because of her mother's insight, but because she herself sought it out until she found it, and dressed herself in it. Her occupation consisted of selling words. She travelled through the country, from the highest and coldest regions to the scorching coast, setting up in markets and fairs four sticks with a cloth awning, under which she protected herself from the sun and the rain to see to her customers. She had no need to call out her wares, because with so much walking here and there everybody knew her. There were those who waited for her year in, year out, and when she appeared in the village with her bundle under her arm they would line up in front of her stall. Her prices were fair. For five cents she would give out poems learnt by heart; for seven she would improve the quality of dreams; for nine she would write love letters; for twelve she would make up insults for irreconcilable enemies. She also sold stories; not fantastical tales but long, true chronicles which she would recite from beginning to end, without skipping a word. That is how she would carry the news from one village to the next. People would pay her to add one or two lines: a child was born, so-and-so died, our children were married, the harvest caught fire. In each place people would sit around her to listen to her when she began to speak, and that is how they would learn about the lives of others, of faraway relatives, the ins and outs of the civil war. To anyone who bought fifty cents' worth she'd give as a gift a secret word to frighten melancholy away. Of course, it wasn't

Isabel Allende was born in Peru in 1942 and lived most of her life in Chile where she was a journalist and playwright.
The turning point of her life, after which her "real writing" began, was the murder of her uncle, President Salvador Allende. Since then she has lived in Venezuela and California. Of the mysteries in her stories, she has said, "The writer asks questions which the reader knows have no answers. It is because of the questions that literature is so powerful. It is because of the questions we read."

the same word for all — that would have been collective deception. Each one received his own, certain that no one, in the entire universe and beyond, would use that word to that specific end.

Belisa Crepusculario was born into a poor family, so poor that they did not even have names to call their children. She came into the world and grew up in the most inhospitable of regions and until her twelfth birthday she had no other virtue or occupation than surviving the hunger and thirst of centuries. During an interminable drought she was forced to bury four younger brothers, and when she understood that her turn had come she decided to set forth across the plains towards the sea, wondering whether on the road she might be able to trick death. She was so stubborn that she succeeded, and not only did she save her own life but also, by chance, she discovered the art of writing. As she reached a village near the coast, the wind dropped at her feet a newspaper page. She lifted the yellow and brittle piece of paper and stood there a long while staring at it, unable to guess its use, until her curiosity overcame her shyness. She approached a man washing his horse in the same muddy waters where she had quenched her thirst.

"What is this?" she asked.

"The sports page of a newspaper," the man answered, not at all surprised at her ignorance.

The answer left the girl astonished, but she didn't want to seem impudent, so she simply enquired about the meaning of the tiny fly legs drawn on the paper.

"They are words, child. Here it says that Fulgencio Barba knocked out Tiznao the Black in the third round."

That day Belisa Crepusculario found out that words fly about loose, with no master, and that anyone with a little cunning can catch them and start a trade. She reflected on her own situation, and realized that, other than becoming a prostitute or a servant in a rich man's kitchen, there were few jobs that she could do. Selling words seemed to her a decent alternative. From that moment she worked at that profession and never took on another. At first she offered her wares without suspecting that words could be written in other places than newspapers. When she became aware of this, she worked out the infinite projections of her business, paid a priest twenty pesos to teach her to read and write, and with the three pesos left over from her savings bought herself a dictionary. She perused it from A to Z, and then threw it into the sea, because she had no intention of swindling her clients with prepackaged words.

One August morning Belisa Crepusculario was sitting under her awning, selling words of justice to an old man who for twenty years had been requesting his pension, when a group of horsemen burst into the marketplace. They were the Colonel's men led by the Mulatto who was known throughout the area for the quickness of his knife and his loyalty towards his chief. Both men, the Colonel and the Mulatto, had spent their lives busy with the civil war, and their names were irremediably linked to calamity and plunder. The warriors arrived in a cloak of noise and dust and as they advanced the terror of a hurricane spread across the marketplace. The chickens escaped in a flutter, the dogs bolted, the women ran away with their children, and there was not a single soul left in the marketplace, except Belisa Crepusculario, who had never before seen the Mulatto, and who was therefore surprised when he addressed her.

"It is you I'm after," he shouted, pointing at her with his rolled-up whip, and even before he finished saying so two men fell upon Belisa Crepusculario, trampling her tent and breaking her inkwell, tied her hands and feet, and lay her like a sailor's sack across the rump of the Mulatto's mount. They set off southwards at a gallop.

Hours later, when Belisa Crepusculario felt she was on the point of dying with her heart turned to sand by the shaking of the horse, she realized that they were stopping and that four powerful hands were placing her on the ground. She tried to stand up and raise her head with dignity, but her strength failed her and she collapsed with a sigh, sinking into a dazzling sleep. She awoke several hours afterwards to the murmurs of night, but she had no time to decipher the sounds because as she opened her eyes she was met by the impatient eyes of the Mulatto, kneeling by her side.

"You're awake at last, woman," he said, offering his canteen for her to drink a sip of firewater mixed with gunpowder to jolt her back to life.

She wanted to know why she had been so mistreated, and he explained that the Colonel required her services. He allowed her to wet her face, and immediately led her to the far end of the camp where the man most feared in the entire land lay resting in a hammock suspended between two trees. She couldn't see his face because it lay hidden in the uncertain shade of the foliage and the ineffaceable shadow of many years living as an outlaw, but she imagined that it must be terrible if the Mulatto addressed him in such humble tones. She was taken aback by his voice, soft and mellifluous like that of a scholar.

"Are you the one who sells words?"

"At your service," she mumbled, screwing up her eyes to see him better in the gloom.

The Colonel rose to his feet and the light in the Mulatto's torch hit him full in the face. She saw his dark skin and his fierce puma eyes, and realized at once that she was facing the loneliest man in the world.

"I want to be President," he said.

He was tired of roaming this cursed land fighting useless wars and suffering defeats that no trick could turn into victories. For many years now he had slept in the open, bitten by mosquitoes, feeding on iguanas and rattlesnake soup, but these minor inconveniences were not reason enough to change his fate. What really bothered him was the terror in other men's eyes. He wanted to enter villages under arcs of triumph, among coloured banners and flowers, with crowds cheering him and bringing him gifts of fresh eggs and freshly-baked bread. He was fed up with seeing how men fled from his approach, how women aborted in fright and children trembled. Because of this, he wanted to become President. The Mulatto had suggested they ride to the Capital and gallop into the Palace to take over the government, in the same way that they took so many other things without asking anyone's permission. But the Colonel had no wish to become simply another dictator, because there had been quite enough of those and anyhow that was no way for him to coax affection from the people. His idea was to be chosen by popular vote in the December elections.

"For that, I need to speak like a candidate. Can you sell me the words for a speech?" the Colonel asked Belisa Crepusculario.

She had accepted many different requests, but none like this one, and yet she felt she could not refuse, for fear that the Mulatto would shoot her between the eyes, or even worse, that the Colonel would burst into tears. Also, she felt an urge to help him, because she became aware of a tingling heat on her skin, a powerful desire to touch this man, go over him with her hands, hold him in her arms.

All night long and much of the following day Belisa Crepusculario rummaged through her stock for words that would suit a presidential oration, while the Mulatto kept a close watch on her, unable to tear his eyes away from her strong walker's legs and untouched breasts. She cast aside words that were dry and harsh, words that were too flowery, too faded through too much use, words that made improbable promises, words lacking truth, muddled words, and was left with only those words that were able to reach unerringly men's thoughts and women's intuitions. Making use of the skills bought from the priest for twenty pesos, she wrote out the speech on a sheet of paper and signalled at the Mulatto for him to untie the rope that held her by the ankles to a tree. She was brought once more into the Colonel's presence, and seeing him again, again she felt the same tingling anxiety she had felt at their first meeting. She gave him the sheet of paper and waited, while he stared at it, holding it with only the tips of his fingers.

"What in Hell's name does it say here?" he asked at last.

"Can't you read?"

"What I know is how to make war," he answered.

She read him the speech out loud. She read it three times, so that he would be able to learn it by heart. When she finished, she saw the emotion drawn on the faces of his soldiers who had assembled to listen to her, and noticed that the Colonel's yellow eyes were shining with excitement, certain that with these words the presidential seat would be his.

"If after hearing it three times, the boys still have their mouths open, then this stuff works, my Colonel," approved the Mulatto.

"How much do I owe you for your job, woman?"

"One peso, my Colonel."

"That's not expensive," he said, opening the pouch which hung from his belt with the leftovers from the latest booty.

"And you also have a right to a bonus. You can have two

secret words, for free," said Belisa Crepusculario.

"How's that?"

She explained to him that for every fifty cents a customer spent, she gave away one word for his exclusive use. The chief shrugged, because he had no interest in her offer, but he didn't want to seem impolite with one who had served him so well. Slowly she approached the leather stool on which he was sitting, and leaned over to give him his two words. Then the man caught a whiff of the mountain smell on her skin, the fiery heat of her hips, the terrible brushing of her hair against him, the wild mint breath whispering into his ear the two secret words that were his by right.

"They're yours, my Colonel," she said drawing back. "You may use them as much as you wish."

The Mulatto escorted Belisa to the edge of the road, never taking his begging mongrel eyes off her, but when he stretched out his hand to touch her she stopped him with a gush of made-up words that dampened his desire, because he believed they were some sort of irrevocable curse.

During the months of September, October, and November the Colonel delivered his speech so many times that, had it not been composed out of long-lasting and glittering words, the constant use would have turned the speech to ashes. He travelled across the country in all directions, riding into cities with a triumphant air, but stopping also in forgotten hamlets where only traces of garbage signalled a human presence, in order to convince the citizens to vote for him. While he spoke, standing on a platform in the middle of the marketplace, the Mulatto and his men distributed sweets and painted his name with scarlet frost on the walls. After the speech was over, the troops would light firecrackers, and when at last they would ride away, a trail of hope remained for many days in midair, like the memory of a comet. Very soon the Colonel became the most popular candidate. It was something never seen before — this man sprung from the civil war, crisscrossed with scars and speaking like a scholar, whose renown spread across the nation moving the very heart of the country. The press began to take notice. Journalists travelled from far to interview him and repeat his words, and the number of his followers grew, as well as the number of his enemies.

"We're doing fine, my Colonel," said the Mulatto after twelve weeks of success.

But the candidate did not hear him. He was repeating his two secret words, as he now did more and more frequently. He would repeat them when nostalgia made him soft, he would murmur them in his sleep, he would carry them with him on horseback, he would think of them just before pronouncing his celebrated speech, and he would surprise himself savouring them at careless moments. And everytime these two words sprang to his mind, he would scent the mountain perfume, the fiery heat, the terrible brushing and the wild mint breath, until he began to roam around like a sleepwalker, and his very own men realized that his life would come to an end before he reached the presidential seat.

"What's happening to you, my Colonel?" the Mulatto asked many times, until one day the chief could bear it no longer and confessed that the two words he carried embedded in his gut were to blame for his mood.

"Tell them to me, to see if that way they lose their power," his faithful assistant begged him.

"I won't, they belong to me alone," the Colonel answered.

Tired of seeing his chief weaken like a man condemned to death, the Mulatto slung his rifle over his shoulder and set off in search of Belisa Crepusculario. He followed her traces throughout the vastness of the land, until he found her under her awning, telling her rosary of tales. He stood in front of her, legs spread out and weapon in hand.

"You're coming with me," he ordered.

She was waiting for him. She picked up her inkwell, folded her awning, draped her shawl over her shoulders and without saying a word climbed on the back of his horse. During the whole journey they never even motioned to one another, because the Mulatto's lust for her had turned to anger and only the fear of her tongue stopped him from whipping her to shreds, as he would have done with anyone else in a similar situation; nor was he willing to tell her how the Colonel went about in a daze, and how what had not been achieved by so many years of battling had suddenly been wrought by a charm whispered in his ear. Two days later they reached the camp, and he immediately took the prisoner to the candidate, in front of all the troops.

"I've brought you this witch so that you can give her back her words, my Colonel, and she give you back your manhood," he said, pointing the rifle's muzzle at the woman's neck.

The Colonel and Belisa Crepusculario gazed at one another for a long moment, measuring one another at a distance. The men then understood that their chief would never be able to rid himself of the charm of those two cursed words, because as they looked on, they saw how the bloodthirsty eyes of the puma softened as the woman now stepped forward and held him by the hand.

Translated from the Spanish by Alberto Manguel

The
BOY'S OWN
Annual

MARGARET ATWOOD

Margaret Atwood was born in Ottawa, Canada in 1939, grew up in Northern Quebec and Ontario and in Toronto. As a reader, she states, "All I want from a good story is what children want.... They are longing to hear a story, but only if you are longing to tell one." She describes her short fictions as "mutations ... not a poem, a short story or a prose poem.... We know what is expected in a given arrangement of words; we know what is supposed to come next. And then it doesn't."

The Boy's Own Annual, 1911, was in my grandfather's attic, along with a pump organ that contained bats, rafter-high piles of Western paperbacks, and a dress form, my grandmother's body frozen in wire when it still had a waist. The attic smelled of dry rot and smoked eels but it had a window, where the sunlight was yellower than anywhere else, because of the dust maybe. This buttery sunlight framed the echoing African caves where the underground streams ran, lightless, haunted by crocodiles, white and eyeless, guarding the entrance to the tunnel carved with Egyptian hieroglyphs and armed with deadly snakes and spiky ambushes planted two thousand years ago to protect the chamber of the sacred pear, which for some reason, in stories like this, was always black. And when the hero snatched it out of the stone forehead looming bulbous and idolatrous there in the darkness, *filthy* was a word they liked, for other religions, the goddess was mad as blazes. Sinister priests with scimitars abounded, they could sniff you out like bloodhounds, their bare feet making no sound, until suddenly there was a set piece and down the hill went everyone, bounding along, loving it, yelling like crazy, bullets thudding into bodies into the scrub, into the surf, onto the waiting ship where Britain stood firm for plunder.

The issue with the last instalment had never come; it wasn't in the attic. So there I was, suspended in mid-story, in 1951, and there I remain, sometime, waiting for the end, or finishing it off myself, in a booklined London study over a stiff brandy, a yarn spun to a few choice gentlemen under the stuffed water buffalo head, a cheerful fire in the grate, or somewhere on the veldt, a bullet in the heart, who can tell where such greedy impulses will lead? Such lust for blind white crocodiles. In those times there were still chiefs in ostrich feathers and enemies worth killing, and loyalty, or so the story said. Through the attic window and its golden dust and flyhusks I could see the barn, unpainted, hay coming out like stuffing from the loft doorway, and around the corner of it my grandmother's cow. She'd hook you if she could, if you didn't have a pitchfork. She was sneaking up on someone invisible; possibly my half uncle, gassed in the first war and never right since. The books had been his once.

The Homecoming

STRANGER

BEI DAO

1

PAPA WAS BACK.

After exactly twenty years of reform through labour, which took him from the North-East to Shanxi, and then from Shanxi to Gansu, he was just like a sailor swept overboard by a wave, struggling blindly against the undertow until miraculously he was tossed by another wave back onto the same deck.

The verdict was: it was entirely a misjudgement, and he has been granted complete rehabilitation. That day, when the leaders of the Theatre Association honoured our humble home to announce the decision, I almost jumped up: when did you become so clever? Didn't the announcement that he was an offender against the people come out of your mouths too? It was Mama's eyes, those calm yet suffering eyes, that stopped me.

Bei Dao (which means "north island") is the pen name of Zhao Zhenkai, who was born in Beijing in 1949. A member of the "lost generation" of China's Cultural Revolution, he coedited the influential literary magazine, Today, *during the days of the Democratic movement.*

Of his writing Dao, considered one of the finest poets in China, says: "Poets should establish through their works a world of their own, a genuine and independent world, an upright world, a world of justice and humanity."

Next came the dress rehearsal for the celebration: we moved from a tiny pigeon-loft into a three-bedroom flat in a big building; sofas, bookcases, desks and chrome folding chairs appeared as if by magic (I kept saying half-jokingly to Mama that these were the troupe's props); relatives and friends came running in and out all day, until the lacquer doorknob was rubbed shiny by their hands, and even those uncles and aunts who hadn't shown up all those years rushed to offer congratulations ... all right, cheer, sing, but what does all this have to do with me? My Papa died a long time ago, he died twenty years ago, just when a little four- or five-year old girl needed a father's love — that's what Mama, the school, kind-hearted souls and the whole social upbringing that starts at birth told me. Not only this, you even wanted me to hate him, curse him, it's even possible you'd have given me a whip so I could lash him viciously! Now it's the other way round, you're wearing a different face. What do you want me to do? Cry, or laugh?

Yesterday at dinner time, Mama was even more considerate than usual, endlessly filling my bowl with food. After the meal, she drew a telegram from the drawer and handed it to me, showing not the slightest sign of any emotion.

"Him?"

"He arrives tomorrow, at 4:50 in the afternoon."

I crumpled the telegram, staring numbly into Mama's eyes.

"Go and meet him, Lanlan." She avoided my gaze.

"I have a class tomorrow afternoon."

"Get someone to take it for you."

I turned towards my room. "I shan't go."

"Lanlan." Mama raised her voice. "He is your father after all!"

"Father?" I muttered, turning away fiercely, as if overcome with fear at the meaning of this word. From an irregular spasm in my heart, I realized it was stitches from the old wound splitting open one by one.

I closed the composition book spread in front of me: Zhang Xiaoxia, 2nd Class, 5th Year. A spirited girl, her head always slightly to one side in a challenging way, just like me as a child. Oh yes, childhood. For all of us life begins with those pale blue copybooks, with those words, sentences and punctuation marks smudged by erasers; or, to put it more precisely, it begins with a certain degree of deception. The teachers delineated life with haloes, but which of them does not turn into a smoke ring or an iron hoop?

Shadows flowed in from the long old-fashioned windows, dulling the bright light on the glass desk-top. The entire staff-room was steeped in drowsy tranquillity. I sighed, tidied my things, locked the door, and crossing the deserted school grounds walked towards home.

The apartment block with its glittering lights was like a huge television screen, the unlit windows composing an elusive image.

After a little while some of the windows lit up, and some went dark again. But the three windows on the seventh floor remained as they were: one bright, two dark. I paced up and down for a long time in the vacant lot piled with white lime and fir poles. On a crooked, broken signboard were the words: "Safety First."

Strange, why is it that in all the world's languages, this particular meaning comes out as the same sound: Papa. Fathers of different colours, temperaments and status all derive the same satisfaction from this sound. Yet I still can't say it. What do I know about him? Except for a few surviving old photographs retaining a childhood dream (perhaps every little girl would have such dreams): him, sitting on an elephant like an Arab sheik, a white cloth wound round his head, a resplendent mat on the elephant's back, golden tassels dangling to the ground ... there were only some plays that once created a sensation and a thick book on dramatic theory which I happened to see at the wastepaper salvage station. What else was there? Yes, add those unlucky letters, as punctual and drab as a clock; stuck in those brown-paper envelopes with their red frames, they were just like death notices suffocating me. I never wrote back, and afterwards, I threw them into the fire without even looking at them. Once, a dear little duckling was printed on a snow-white envelope, but when I tore it open and looked, I was utterly crushed. I was so upset I cursed all ugly ducklings, counting up their vices one by one: greed, pettiness, slovenliness ... because they hadn't brought me good luck. But what luck did I deserve?

The lift had already closed for the day, and I had to climb all the way up. I stopped outside the door to our place and listened, holding my breath. From inside came the sounds of the television hum and the clichés of an old film. God, give me courage!

As soon as I opened the door, I heard my younger brother's gruff voice: "Sis's back." He rushed up as if making an assault on the enemy, helping me to take off my coat. He was almost twenty, but still full of a childish attachment to me, probably because I had given him the maternal love which had seemed too heavy a burden for Mama in those years.

The corridor was very dark and the light from the kitchen split the darkness into two. He was standing in the doorway of the room opposite, standing in the other half of darkness, and next to him was Mama. The reflection from the television screen flickered behind their shoulders.

A moment of dead silence.

Finally, he walked over, across the river of light. The light, the deathly-white light, slipped swiftly over his wrinkled and mottled neck and face. I was struck dumb: was this shrivelled little old man him? Father. I leant weakly against the door.

He hesitated a moment and put out his hand. My small hand disappeared in his stiff, big-jointed hand. These hands didn't match his body at all.

"Lanlan." His voice was very low, and trembled a little.

Silence.

"Lanlan," he said again, his voice becoming a little more positive, as if he were waiting eagerly for something.

But what could I say?

"You're back very late. Have you had dinner?" said Mama.

"Mm." My voice was so weak.

"Why is everyone standing? Come inside," said Mama.

He took me by the hand. I followed obediently. Mama turned on the light and switched off the television with a click. We sat down on the sofa. He was still clutching my hand tightly, staring at me intently. I evaded his eyes, and let my gaze fall on the blow-up plastic doll on the window sill.

An unbearable silence.

"Lanlan," he called once again.

I was really afraid the doll might explode, sending brightly-coloured fragments flying all over the room.

"Have you had your dinner?"

I nodded vigorously.

"Is it cold outside?"

"No." Everything was so normal. The doll wouldn't burst. Perhaps it would fly away suddenly like a hydrogen balloon, out the window, above the houses full of voices, light and warmth, and go off to search for the stars and moon.

"Lanlan." His voice was full of compassion and pleading.

All of a sudden, my just-established confidence swiftly collapsed. I felt a spasm of alarm. Blood pounded at my temples. Fiercely I pulled back my hand, rushed out the door into my own room and flung myself head-first onto the bed. I really felt like bursting into tears.

The door opened softly; it was Mama. She came up to the bed, sat down in the darkness and stroked my head, neck and shoulders. Involuntarily my whole body began to tremble as if with cold.

"Don't cry, Lanlan."

Cry? Mama, if I could still cry the tears would surely be red, they'd be blood.

She patted me on the back. "Go to sleep, Lanlan, everything will pass."

Mama left.

Everything will pass. Huh, it's so easily said, but can twenty years be written off at one stroke? People are not reeds, or leaches, but oysters, and the sands of memory will flow with time to change into a part of the body itself, teardrops that will never run dry.

... a basement. Mosquitoes thudded against the searing light bulb. An old man covered with cuts and bruises was tied up on the pommell horse, his head bowed, moaning hoarsely. I lay in the corner sobbing. My knees were cut to ribbons by the broken glass; blood and mud mixed together ...

I was then only about twelve years old. One night, when Mama couldn't sleep, she suddenly hugged me and told me that Papa was a good man who had been wrongly accused. At these words hope flared up in the child's heart: for the first time she might be able to enjoy the same rights as other children. So I ran all around, to the school, the Theatre Association, the neighbourhood committee and the Red Guard headquarters, to prove Papa's innocence to them. Disaster was upon us, and those louts took me home savagely for investigation. I didn't know what was wrong with Mama, but she repudiated all her words in front of her daughter. All the blame fell on my small shoulders. Mama repented, begged, wished herself dead, but what was the use? I was given heavy labour and punished by being made to kneel on broken glass.

... the old man raised his bloody face: "Give me some water, water, water?" Staring with frightened eyes, I forgot the pain, huddling tightly into the corner. When dawn came and the old man breathed his last, I fainted with fright too. The blood congealed on my knees ...

Can I blame Mama for this?

2

The sky was so blue it dazzled the eyes, its intense reflections shining on the ground. My hair tied up in a ribbon, I was holding a small empty bamboo basket and standing amidst the dense waist-high grass. Suddenly from the jungle opposite appeared an elephant, the tassels of the mat on its back dangling to the ground; Papa sat proudly on top, a white turban on his head. The elephant's trunk waved to and fro, and with a snort it curled around me and placed me up in front of Papa. We marched forward, across the coconut grove streaked with leaping sunlight, across the hills and gullies gurgling with springs. I suddenly turned my head and cried out in alarm. A little old man was sitting behind me, his face blurred with blood; he was wearing convict clothes and on his chest were printed the words "Reform Through Labour." He was moaning hoarsely, "give me some water, water, water ..."

I woke up in fright.

It was five o'clock, and outside it was still dark. I stretched out my hand and pulled out the drawer of the bedside cupboard, fumbled for cigarettes and lit one. I drew back fiercely and felt more relaxed. The white cloud of smoke spread through the darkness and finally floated out through the small open-shuttered window. The glow from the cigarette alternately brightened and dimmed as I strained to see clearly into the depths of my heart, but other than the ubiquitous silence, the relaxation induced by the cigarette and the vague emptiness left by the nightmare, there was nothing.

I switched on the desk lamp, put on my clothes and opened the door quietly. There was a light on in the kitchen and a rustling noise. Who was up so early? Who?

Under the light, wearing a black cotton-padded vest, he was crouching over the waste-paper basket with his back towards me, meticulously picking though everything; spread out beside him were such spoils as vegetable leaves, trimmings and fish heads.

I coughed.

He jumped and looked round in alarm, his face deathly-white, gazing in panic towards me.

The fluorescent light hummed.

He stood up slowly, one hand behind his back, making an effort to smile. "Lanlan, I woke you up."

"What are you doing?"

"Oh, nothing, nothing." He was flustered, and kept wiping his trousers with his free hand.

I put out my hand. "Let me see."

After some hesitation he handed the thing over. It was just an ordinary cigarette packet, with nothing odd about it except that it was soiled in one corner.

I lifted my head, staring at him in bewilderment.

"Oh, Lanlan," beads of sweat started from his balding head, "yesterday I forgot to examine this cigarette packet when I threw it away, just in case I wrote something on it; it would be terrible if the team leader saw it."

"Team leader?" I was even more baffled. "Who's the team leader?"

"The people who oversee us prisoners are called team leaders." He fished out a handkerchief and wiped the sweat away. "Of course, I know, it's beyond their reach, but better to find it just in case ..."

My head began to buzz. "All right, that's enough."

He closed his mouth tightly, as if he had even bitten out his tongue. I really hadn't expected our conversation would begin like this. For the first time I looked carefully at him. He seemed even older and paler than yesterday, with a short greyish stubble over his sunken cheeks, wrinkles that seemed to have been carved by a knife around his lacklustre eyes and an ugly sarcoma on the tip of his right ear. I could not help feeling some compassion for him.

"Was it very hard there?"

"It was all right, you get used to it."

Get used to it! A cold shiver passed through me. Dignity. Wire netting. Guns. Hurried footsteps. Dejected ranks. Death. I crumpled up the cigarette packet and tossed it into the waste-paper basket. "Go back to sleep, it's still early."

"I've had enough sleep, reveille's at 5:30." He turned to tidy up the scattered rubbish.

Back in my room, I pressed my face against the ice-cold wall. It was quite unbearable, to begin like this, what should I do next? Wasn't he a man of great integrity before? Ah, Hand of Time, you're so cruel and indifferent, to knead a man like putty, you destroyed him before his daughter could remember her father's real face clearly ... eventually I calmed down, packed my things into my bag and put on my overcoat.

Passing through the kitchen, I came to a standstill. He was at the sink, scrubbing his big hands with a small brush, the green soap froth dripping down like sap.

"I'm going to work."

"So early?" He was so absorbed he did not even raise his head.

"I'm used to it."

I did not turn on the light, going down along the darkness, along each flight of stairs.

3

For several days in a row I came home very late. When Mama asked why, I always offered the excuse that I was busy at school. As soon as I got home, I would dodge into the kitchen and hurriedly rake up a few leftovers, then bore straight into my own little nest. I seldom ran into him, and even when we did meet I would hardly say a word. Yet it seemed his silence contained enormous compunction, as if to apologize for that morning, for his unexpected arrival, for my unhappy childhood, these twenty years and my whole life.

My brother was always running in like a spy to report on the situation, saying things like "He's planted a pot of peculiar dried-up herbs," "All afternoon he stared at the fish in the tank," "He's burned a note again" ... I would listen without any reaction. As far as I was concerned, it was all just a continuation of that morning, not worth making a fuss about. What was strange was my brother, talking about such things so flatly, not tinged by any emotion at all, not feeling any heavy burden on his mind. It was not surprising; by the time he was born Papa had already flown away, and besides, in those years he was brought up in Grandma's house, and with Mama's wings and mine in turn hanging over

Grandma's little window as well, he never saw the ominous sky.

One evening, as I was lying on the bed smoking, someone knocked at the door. I hurriedly stuffed the cigarette butt in a small tin box, as Mama came in.

"Smoking again, Lanlan?"

As if nothing had happened I turned over the pages of a novel beside my pillow.

"The place smells of smoke, open a window."

Thank heavens, she hadn't come to nag. But then I realized that there was something strange in her manner. She sat down beside the small desk, absently picked up the ceramic camel pen-rack and examined it for a moment before returning it to its original place. How would one put it in diplomatic language? Talks, yes, formal talks ...

"Lanlan, you're not a child anymore." Mama was weighing her words.

It had started; I listened with respectful attention.

"I know you've resented me since you were little, and you've also resented him and everyone else in the world, because you've had enough suffering ... but Lanlan, it isn't only you who's suffered."

"Yes, Mama."

"When you marry Jianping, and have children, you'll understand a mother's suffering ..."

"We don't want children if we can't be responsible for their future."

"You're not blaming us, Lanlan," Mama said painfully.

"No, not blaming. I'm grateful to you, Mama, it wasn't easy for you in those years ..."

"Do you think it was easy for him?"

"Him?" I paused. "I don't know, and I don't want to know either. As a person, I respect his past ..."

"Don't you respect his present? You should realize, Lanlan, his staying alive required great courage!"

"That's not the problem, Mama. You say this because you lived together for many years, but I, I can't make a false display of affection ..."

"What are you saying!" Mama grew angry and raised her voice. "At least one should fulfil one's own duties and obligations!"

"Duties? Obligations?" I started to laugh, but it was more painful than crying. "I heard a lot about them during those years. I don't want to lose any more, Mama."

"But what have you gained?"

"The truth."

"It's a cold and unfeeling truth!"

"I can't help it," I spread out my hands, "that's how life is."

"You're too selfish!" Mama struck the desk with her hand and got up, the loose flesh on her face trembling. She stared furiously at me for a moment, then left, shutting the door heavily.

Selfish, I admit it. In those years, selfishness was a kind of instinct, a means of self-defence. What could I rely on except this? Perhaps I shouldn't have provoked Mama's anger, perhaps I should really be a good girl and love Papa, Mama, my brother, life, and myself.

4

During the break between classes, I went into the reception office and rang Jiangping.

"Hello, Jianping, come over this evening."

"What's up? Lanlan?" he was shouting over the clatter of the machines sounding hoarse and weary.

"He's back."

"Who? Your father?"

"Clever one, come over and help; it's an absolutely awful situation."

He started to laugh.

"Huh, if you laugh, just watch out!" I clenched my fists and banged down the receiver.

It's true, Jianping has the ability to head off disaster. The year when the production brigade chief withheld the grain ration from us educated youth, it was he who led the whole bunch of us to snatch it all back. Although I normally appear to be quite sharpwitted, I always have to hide behind his broad shoulders whenever there's a crisis.

That afternoon I had no classes and hurried home early. Mama had left a note on the table, saying that she and Papa had gone to call on some old friends and would eat when they returned. I kneaded some dough, minced the meat filling and got everything ready to wrap the dumplings.

Jianping arrived. He brought with him a breath of freshness and cold, his cheeks flushed red, brimming with healthy vitality. I snuggled up against him at once, my cheek pressed against the cold buttons on his chest, like a child who feels wronged but has nowhere to pour out her woes. I didn't say anything, what could I say?

We kissed and hugged for a while, then sat down and wrapped dumplings, talking and joking as we worked. From gratitude, relaxation and the vast sleepiness that follows affection, I was almost on the verge of tears.

When my brother returned, he threw off his work clothes, drank a mouthful of water, and flew off like a whirlwind.

It was nearly eight when they got home. As they came in, it gave them quite a shock to see us. Mama could not conceal a conciliatory and motherly smile of victory; Papa's expression was much more complicated. Apart from the apologetic look of the last few days, he also seemed to feel an irrepressible pleasure at this surprise, as well as a precautionary fear.

"This is Jianping, this is ..." My face was suffocated with red.

"This is Lanlan's father," Mama filled in.

Jianping held out his hand and boomed, "How do you do, Uncle!"

Papa grasped Jianping's hand, his lips trembling for a long time. "So you're, so you're Jianping, fine, fine ..."

Delivering the appropriate courtesies, Jianping gave the old man such happiness he was at a loss what to do. It was quite clear to me that his happiness had nothing to do with those remarks, but was because he felt that at last he'd found a bridge between him and me, a strong and reliable bridge.

At dinner, everyone seemed to be on very friendly terms, or at least that's how it appeared on the surface. Several awkward silences were covered over by Jianping's jokes. His conversation was so witty and lively that it even took me by surprise.

After dinner, Papa took out his Zhonghua* cigarettes from a tin cigarette case to offer to Jianping. This set them talking about the English method of drying tobacco and moving on to soil salinization, insect pests among peanuts and vine-grafting. I sat bolt upright beside them, smiling like a mannequin in a shop window.

Suddenly, my smile began to vanish. Surely this was a scene from a play? Jianping was the protagonist — a clever son-in-law, while I, I was the meek and mild new bride. For reasons only the devil could tell, everyone was acting to the hilt, striving to forget something in this scene. Acting happiness, acting calmness, acting glossed-over suffering. I suddenly felt that Jianping was an outsider to the fragmented, shattered suffering of this family.

I began to consider Jianping in a different light. His tone, his gestures, even his appearance, all had an unfamiliar flavour. This wasn't real, this wasn't the old he. Could strangeness be contagious? How frightening.

Jianping hastily threw me an enquiring glance, as if expecting me to repay the role he was playing with a commending smile. This made me feel even more disgusted. I was disgusted with him, and with myself, disgusted with everything the world is made of, happiness and sorrow, reality and sham, good and evil.

Guessing this, he wound up the conversation. He looked at his watch, said a few thoroughly polite bits of nonsense, and got to his feet.

As usual, I accompanied him to the bus-stop. But along the way, I said not a single word, keeping a fair distance from him. He dejectedly thrust his hands in his pockets, kicking a stone.

An apartment block ahead hid the night. I felt alone. I longed to know how human beings survive behind those countless containers of suffering broken families. Yet in these containers, memory is too frightening. It can only deepen the suffering and divide every family until everything turns to powder.

When we reached the bus-stop, he stood with his back to me, gazing at the distant lights. "Lanlan, do I still need to explain?"

"There's no need."

He leapt onto the bus. Its red tail-lights flickering, it disappeared round the corner.

5

Today there was a sports meeting at the school, but I didn't feel like it at all. Yesterday afternoon, Zhang Xiaoxia kept pestering me to come and watch her in the 100 metre race. I just smiled, without promising anything. She pursed her little mouth and, fanning her cheeks, which were streaming with sweat, with her handkerchief, stared out the window in a huff. I put my hands on her shoulders and turned her round. "I'll go then, all right?" Her face broadening into dimples, she struggled free of me in embarrassment and ran off. How easy it is to deceive a child.

I stretched, and started to get dressed. The winter sunlight seeped through the fogged-up window, making everything seem dim and quiet, like an extension of sleep and dreams. When I came out of my room, it was quiet and still; evidently everyone had gone out. I washed my hair and put my washing to soak, dashing busily to and fro. When everything was done, I sat down to eat breakfast. Suddenly I sensed that someone was standing behind me, and when I looked round it was Papa, standing stiffly in the kitchen doorway and staring at me blankly.

"Didn't you go out?" I asked.

"Oh, no, no, I was on the balcony. You're not going to school today?"

"No. What is it?"

"I thought," he hesitated, "we might go for a walk in the park, what do you think?" There was an imploring note in his voice.

"All right." Although I didn't turn around, I could feel that his eyes had brightened.

It was a warm day, but the morning mist had still not faded altogether, lingering around eaves and treetops. Along the way, we said almost nothing. But when we entered the park, he pointed at the tall white poplars by the side of the road. "The last time I brought you here, they'd just been planted." But I didn't remember it at all.

After walking along the avenue for a while, we sat down on a bench beside the lake. On the cement platform in front of us, several old wooden boats, corroded by wind and rain, were lying upside down, dirt and dry leaves forming a layer over them. The ice on the surface of the water crackled from time to time.

He lit a cigarette.

"Those same boats," he said pensively.

"Oh?"

"They're still the same boats. You used to like sitting in the stern, splashing with your bare feet and shouting, 'Motor-boat! Motor-boat!'" The shred of a smile of memory appeared on his face. "Everyone said you were like a boy ..."

"Really?"

"You liked swords and guns; whenever you went into a toyshop you'd always want to come out with a whole array of weapons."

"Because I didn't know what they were used for."

All at once, a shadow covered his face and his eyes darkened. "You were still a child then ..."

Silence, a long silence. The boats lying on the bank were turned upside down here. They were covering a little girl's silly cries, a father's carefree smile, softdrink bottle-tops, a blue satin ribbon, children's books and toy guns, the taste of earth in the four seasons, the passage of twenty years ...

"Lanlan," he said suddenly, his voice very low and trembling, "I, I beg your pardon."

My whole body began to quiver.

"When your mother spoke of your life in these years, it was as if my heart was cut with a knife. What is a child guilty of?" His hand clutched at the air and came to rest against his chest.

"Don't talk about these things," I said quietly.

"To tell you the truth, it was for you that I lived in those years. I thought if I paid for my crime myself, perhaps life would be a bit better for my child, but ... he choked with sobs, "you can blame me, Lanlan, I didn't have the ability to protect you, I'm not worthy to be your father..."

"No, don't, don't ..." I was trembling, my whole body went weak, all I could do was shake my hands. How selfish I was! I thought only of myself, immersed myself only in my own sufferings, even making suffering a kind of pleasure and a wall of defence against others. But how did he live? For you, for your selfishness, for your heartlessness! Can the call of blood be so feeble? Can what is called human nature have completely died out in my heart?

"... twenty years ago, the day I left the house, it was a Sunday. I took an afternoon train, but I left at dawn; I didn't want you to remember this scene. Standing by your little bed, the tears streaming down, I thought to myself: 'Little Lanlan, shall we ever meet again?' You were sleeping so soundly and sweetly, with your little round dimples ... the evening before as you were going to bed, you hugged my neck and said in a soft voice, 'Papa, will you take me out tomorrow?' 'Papa's busy tomorrow.' You went into a sulk and pouted unhappily. I had to promise. Then you asked again, 'Can we go rowing?' 'Yes, we'll go rowing.' And so you went to sleep quite satisfied. But I deceived you, Lanlan, when you woke up the next day, what could you think..."

"Papa!" I blurted out, flinging myself on his shoulder and crying bitterly.

With trembling hands he stroked my head. "Lanlan, my child."

"Forgive me, Papa," I said, choked with sobs. "I'm still your little Lanlan, always ..."

"My little Lanlan, always."

A bird whose name I don't know hovered over the lake, crying strangely, adding an even deeper layer of desolation to this bleak winter scene.

I lay crying against Papa's shoulder for a long time. My tears seeped drop by drop into the coarse wool of his overcoat. I seemed to smell the pungent scent of tobacco mingling with the smell of mud and sweat. I seemed to see him in the breaks between heavy labour, leaning wearily against the pile of dirt and rolling a cigarette, staring into the distance through the fork between the guard's legs. He was digging the earth shovelful after shovelful, straining himself to fling it towards the pit side. The guard's legs. He was carrying his bowl, greedily draining the last mouthful of vegetable soup. The guard's legs ... I dared not think any more, I dared not. My powers of imagining suffering were limited after all. But he actually lived in a place beyond the powers of human imagination. Minute after minute, day after day, oh God, a full twenty years ... no, amidst suffering, people should be in communication with one another, suffering can link people's souls even more than happiness, even if the soul is already numb, already exhausted ...

"Lanlan, look," he drew a beautiful necklace from his pocket, "I made this just before I left there from old toothbrush handles. I wanted to give you a kind of present, but then I was afraid you wouldn't want this crude toy ..."

"No, I like it." I took the necklace, moving the beads lightly to and fro with my fingers, each of these wounded hearts ...

On the way back, Papa suddenly bent over and picked up a piece of paper, turning it over and over in his hand. Impulsively I pulled up his arm and laid my head on his shoulder. In my heart I understood that this was because of a new strangeness, and an attempt to resist this strangeness.

Here on this avenue, I seemed to see a scene from twenty years earlier. A little girl with a blue ribbon in her hair, both fists outstretched, totters along the edge of the concrete road. Beside her walks a middle-aged man relaxed and at ease. A row of little newly-planted poplars separates them. And these little trees, as they swiftly swell and spread, change into a row of huge insurmountable bars. Symbolizing this are twenty years of irregular growth rings.

"Papa, let's go."

He tossed away the piece of paper and wiped his hand carefully on his handkerchief. We walked on again.

Suddenly I thought of Zhang Xiaoxia. At this moment, she'll actually be in the race. Behind rises a puff of white smoke from the starting gun and amid countless faces and shrill cries falling away behind her, she dashes against the white finishing tape.

Zhonghua: a trademark of one of the best cigarettes in China.

EVERYTHING

And Nothing

JORGE LUIS BORGES

THERE WAS NO ONE IN HIM; BEHIND HIS face (which even in the poor paintings of the period is unlike any other) and his words, which were copious, imaginative, and emotional, there was nothing but a little chill, a dream not dreamed by anyone. At first he thought everyone was like him, but the puzzled look on a friend's face when he remarked on that emptiness told him he was mistaken and convinced him forever that an individual must not differ from his species. Occasionally he thought he would find in books the cure for his ill, and so he learned the small Latin and less Greek of

Jorge Luis Borges was born in 1899 in Argentina where he learned English before Spanish, was educated in Switzerland, and returned to live in Buenos Aires where he served as Director of the National Library for a time and a professor of English Literature. Describing one of his favourite writers, he defines himself: "{Joseph} Conrad thought that when one wrote, even in a realistic way, about the world, one was writing a fantastic story because the world itself is fantastic and unfathomable and mysterious." He died in 1986.

which a contemporary was to speak. Later he thought that in the exercise of an elemental human rite he might well find what he sought, and he let himself be initiated by Anne Hathaway one long June afternoon. At twenty-odd he went to London. Instinctively, he had already trained himself in the habit of pretending that he was someone, so it would not be discovered that he was no one. In London he hit upon the profession to which he was predestined, that of the actor, who plays on stage at being someone else. His playacting taught him a singular happiness, perhaps the first he had known; but when the last line was applauded and the last corpse removed from the stage, the hated sense of unreality came over him again. He ceased to be Ferex or Tamburlaine and again became a nobody. Trapped, he fell to imagining other heroes and other tragic tales. Thus, while in London's bawdyhouses and taverns his body fulfilled its destiny as body, the soul that dwelled in it was Caesar, failing to heed the augurer's admonition, and Juliet, detesting the lark, and Macbeth, conversing on the heath with the witches, who are also the fates. Nobody was ever as many men as that man, who like the Egyptian Proteus managed to exhaust all the possible shapes of being. At times he slipped into some corner of his work a confession, certain that it would not be deciphered; Richard affirms that in his single person he plays many parts, and Iago says with strange words, "I am not what I am." His passages on the fundamental identity of existing, dreaming, and acting are famous.

Twenty years he persisted in that controlled hallucination, but one morning he was overcome by the surfeit and the horror of being so many kings who die by the sword and so many unhappy lovers who converge, diverge, and melodiously agonize. That same day he disposed of his theatre. Before a week was out he had returned to the village of his birth, where he recovered the trees and the river of his childhood; and he did not bind them to those others his muse had celebrated, those made illustrious by mythological allusions and Latin phrases. He had to be someone; he became a retired impresario who has made his fortune and who interests himself in loans, lawsuits, and petty usury. In this character he dictated the arid final will and testament that we know, deliberately excluding from it every trace of emotion and of literature. Friends from London used to visit his retreat, and for them he would take on again the role of poet.

The story goes that, before or after he died, he found himself before God and he said: "I, who have been so many men in vain, want to be one man: myself." The voice of God replied from a whirlwind: "Neither am I one self; I dreamed the world as you dreamed your work, my Shakespeare, and among the shapes of my dream are you, who, like me, are many persons — and none."

Translated by Mildred Boyer and Harold Morland

A Commencement
ADDRESS
(Williams College, 1984)

JOSEPH BRODSKY

LADIES AND GENTLEMEN
OF THE CLASS OF 1984:

Joseph Brodsky was born in Leningrad in 1940, was sentenced to five years of hard labour in the Arkhangelsk region, was exiled involuntarily in 1972, and won the Nobel Prize for Literature in 1987. He writes: *"If what distinguishes us from other species is speech, then poetry, which is the supreme linguistic operation, is our anthropological, indeed genetic, goal. Anyone who regards poetry as an entertainment … commits an anthropological crime, in the first place, against himself."*

No matter how daring or cautious you may choose to be, in the course of your life you are bound to come into direct physical contact with what's known as Evil. I mean here not a property of the gothic novel but, to say the least, a palpable social reality that you in no way can control. No amount of good nature or cunning calculations will prevent this encounter. In fact, the more calculating, the more cautious you are, the greater is the likelihood of this rendezvous, the harder its impact. Such is the structure of life that what we regard as Evil is capable of a fairly ubiquitous presence if only because it tends to appear in the guise of good. You never see it crossing your threshold announcing itself: "Hi, I'm Evil!" That, of course, indicates its secondary nature, but the comfort one may derive from this observation gets dulled by its frequency.

A prudent thing to do, therefore, would be to subject your notions of good to the closest possible scrutiny, to go, so to speak, through your entire wardrobe checking which of your clothes may fit a stranger. That, of course, may turn into a full-time occupation, and well it should. You'll be surprised how many things you considered your own and good can easily fit, without much adjustment, your enemy. You may even start to wonder whether he is not your mirror image, for the most interesting thing about Evil is that it is wholly human.

To put it mildly, nothing can be turned and worn inside out with greater ease than one's notion of social justice, civic conscience, a better future, etc. One of the surest signs of danger here is the number of those who share your views, not so much because unanimity has the knack of degenerating into uniformity as because of the probability — implicit in great numbers — that noble sentiment is being faked.

By the same token, the surest defence against Evil is extreme individualism, originality of thinking, whimsicality, even — if you will — eccentricity. That is, something that can't be feigned, faked, imitated; something even a seasoned imposter couldn't be happy with. Something, in other words, that can't be shared, like your own skin: not even by a minority. Evil is a sucker for solidity. It always goes for big numbers, for confident granite, for ideological purity, for drilled armies and balanced sheets. Its proclivity for such things has to do presumably with its innate insecurity, but this realization, again, is of small comfort when Evil triumphs.

Which it does: in so many parts of the world and inside ourselves. Given its volume and intensity, given, especially, the fatigue of those who oppose it, Evil today may be regarded not as an ethical category but as a physical phenomenon no longer measured in particles but mapped geographically. Therefore the reason I am talking to you about all this has nothing to do with your being young, fresh, and facing a clean slate. No, the slate is dark with dirt and it is hard to believe in either your ability or your will to clean it. The purpose of my talk is simply to suggest to you a mode of resistance which may come in handy to you one day; a mode that may help you to emerge from the encounter with Evil perhaps less soiled, if not necessarily more triumphant than your precursors. What I have in mind, of course, is the famous business of turning the other cheek.

I assume that one way or another you have heard about the interpretations of this verse from the Sermon on the Mount by Leo Tolstoy, Mahatma Gandhi, Martin Luther King, Jr., and many others. In other words, I assume that you are familiar with the concept of nonviolent, or passive, resistance, whose main principle is returning good for evil, that is, not responding in kind. The fact that the world today is what it is suggests, to say the least, that this concept is far from being cherished universally. The reasons for its unpopularity are twofold. First, what is required for this concept to be put into effect is a margin of democracy. That is precisely what 86 percent of the globe lacks. Second, it is common sense that tells a victim that his only gain in turning the other cheek and not responding in kind yields, at best, a moral victory, i.e., something quite immaterial. The natural reluctance to expose yet another part of your body to a blow is justified by a suspicion that this sort of conduct only agitates and enhances Evil; that a moral victory can be mistaken by the adversary for his impunity.

There are other, graver reasons to be suspicious. If the first blow hasn't knocked all the wits out of the victim's head, he may realize that turning the other cheek amounts to manipulation of the offender's sense of guilt, not to speak of his karma. The moral victory itself may not be so moral after all, not only because suffering often has a narcissistic aspect to it, but also because it renders the victim superior, that is, better than his enemy. Yet no matter how evil your enemy is, the crucial thing is that he is human; and although incapable of loving another like ourselves, we nonetheless know that Evil takes root when one man starts to think that he is better than another. (This is why you've been hit on your right cheek in the first place.) At best, therefore, what one can get from turning the other cheek to one's enemy is the satisfaction of alerting the latter to the futility of his action. "Look," the other cheek says, "what you are hitting is just flesh. It's not me. You can't crush my soul." The trouble, of course, with this kind of attitude is that the enemy may just accept the challenge.

Twenty years ago the following scene took place in one of the numerous prison yards of northern Russia. At seven o'clock in the morning the door of a cell was flung open and on its threshold stood a prison guard, who addressed its inmates: "Citizens! The collective of this prison's guards challenge you, the inmates, to socialist competition in chopping the lumber amassed in our yard." In those parts there is no central heating, and the local police, in a manner of speaking, tax all the nearby lumber companies for one-tenth of their produce. By the time I am describing, the prison yard looked like a veritable lumberyard: the piles were two to three stories high, dwarfing the one-storied quadrangle of the prison itself. The need for chopping was evident, although socialist competitions of this sort had happened before. "And what if I refuse to take part in this?" inquired one of the inmates. "Well, in that case no meals for you," replied the guard.

Then axes were issued to inmates, and the cutting started. Both prisoners and guards worked in earnest, and by noon all of them, especially the always underfed prisoners, were exhausted. A break was announced and people sat down to eat: except the fellow who asked the question. He kept swinging his axe. Both prisoners and guards exchanged jokes about him, something about Jews being normally regarded as smart people whereas this man ... and so forth. After the break they resumed the work, although in a somewhat more flagging manner. By four o'clock the guards quit, since for them it was the end of their shift; a bit later the inmates stopped too. The man's axe still kept swinging. Several times he was urged to stop, by both parties, but he paid no attention. It seemed as though he had acquired a certain rhythm he was unwilling to break; or was it a rhythm that possessed him?

To the others, he looked like an automaton. By five o'clock, by six o'clock, the axe was still going up and down. Both guards and inmates were now watching him keenly, and the sardonic expression on their faces gradually gave way first to one of bewilderment and then to one of terror. By seven-thirty the man stopped, staggered into his cell, and fell asleep. For the rest of his stay in that prison, no call for socialist competition between guards and inmates was issued again, although the wood kept piling up.

I suppose the fellow could do this — twelve hours of straight chopping — because at the time he was quite young. In fact, he was then twenty-four. Only a little older than you are. However, I think there could have been another reason for his behaviour that day. It's quite possible that the young man — precisely because he was young — remembered the text of the Sermon on the Mount better than Tolstoy and Gandhi did. Because the Son of Man was in the habit of speaking in triads, the young man could have recalled that the relevant verse doesn't stop at

but whosoever shall smite thee on thy right cheek,
turn to him the other also

but continues without either period or coma:

And if any man will sue thee at the law, and take
away thy coat, let him have thy cloak also.
And whosoever shall compel thee to go a mile,
go with him twain.

Quoted in full, these verses have in fact very little to do with nonviolent or passive resistance, with the principles of not responding in kind and returning good for evil. The meaning of these lines is anything but passive, for its suggests that evil can be made absurd through excess; it suggests rending evil absurd through dwarfing its demands with the volume of your compliance, which devalues the harm. This sort of thing puts a victim into a very active position, into the position of a mental aggressor. The victory that is possible here is not a moral but an existential one. The other cheek here sets in motion not the enemy's sense of guilt (which he is perfectly capable of quelling) but exposes his sense and faculties to the meaninglessness of the whole enterprise: the way every form of mass production does.

Let me remind you that we are not talking here about a situation involving a fair fight. We are talking about situations where one finds oneself in a hopelessly inferior position from the very outset, where one has no chance of fighting back, where the odds are overwhelmingly against one. In other words, we are talking about the very dark hours in one's life, when one's sense of moral superiority over the enemy offers no solace, when this enemy is too far gone to be shamed or made nostalgic for abandoned scruples, when one has at one's disposal only one's face, coat, cloak, and a pair of feet that are still capable of walking a mile or two.

In this situation there is very little room for tactical manoeuvre. So turning the other cheek should be your conscious, cold, deliberate decision. Your chances of winning, however dismal they are, all depend on whether or not you know what you are doing. Thrusting forward your face with the cheek toward the enemy, you should know that this is just the beginning of your ordeal as well as that of the verse — and you should be able to see yourself through the entire sequence, through all three verses from the Sermon on the Mount. Otherwise, a line taken out of context will leave you crippled.

To base ethics on a faultily quoted verse is to invite doom, or else to end up becoming a mental bourgeois enjoying the ultimate comfort: that of his convictions. In either case (of which the latter with its membership in well-intentioned movements and nonprofit organizations is the least palatable) it results in yielding ground to Evil, in delaying the comprehension of its weaknesses. For Evil, may I remind you, is only human.

Ethics based on this faultily quoted verse have changed nothing in post-Gandhi India, save the colour of its administration. From a hungry man's point of view though, it's all the same who makes him hungry. I submit that he may even prefer a white man to be responsible for his sorry state if only because this way social evil may appear to come from elsewhere and may perhaps be less efficient than the suffering at the hand of his own kind. With an alien in charge, there is still room for hope, for fantasy.

Similarly in post-Tolstoy Russia, ethics based on this misquoted verse undermined a great deal of the nation's resolve in confronting the police state. What has followed is known all too well: six decades of turning the other cheek transformed the face of the nation into one big bruise, so that the state today, weary of its violence, simply spits at that face. As well as the face of the world. In other words, if you want to secularize Christianity, if you want to translate Christ's teachings into political terms, you need something more than modern political mumbo-jumbo: you need to have the original — in your mind at least if it hasn't found room in your heart. Since He was less a good man than a divine spirit, it's fatal to harp on His goodness at the expense of His metaphysics.

I must admit that I feel somewhat uneasy talking about these things: because turning or not turning that other cheek, is after all, an extremely intimate affair. The encounter always occurs on a one-to-one basis. It's always your skin, your coat and cloak, and it is your limbs that will have to do the walking. To advise, let alone to urge, anyone about the use of these properties is, if not entirely wrong, indecent. All I aspire to do here is to erase from your minds a cliché that harmed so many and yielded so little. I also would like to instill in you the idea that as long as you have your skin, coat, cloak, and limbs, you are not yet defeated, whatever the odds are.

There is, however, a greater reason for one to feel uneasy about discussing these matters in public; and it's not only your own natural reluctance to regard your young selves as potential victims. No, it's rather mere sobriety, which makes one anticipate among you potential villains as well, and it is a bad strategy to divulge the secrets of resistance in front of the potential enemy. What perhaps relieves one from a charge of treason or, worse still, of projecting the tactical status quo into the future, is the hope that the victim will always be more inventive, more original in his thinking, more enterprising than the villain. Hence the chance that the victim may triumph.

A
LETTER
to our son

PETER CAREY

Peter Carey was born in Bacchus Marsh, Australia in 1943 and won the Booker Prize in 1988. Considering the difference between self-expression and art, he observes, "When I think about my earlier stories, I was moved by them but I understand why readers weren't. They were written with a rather removed narrative tone; I realized I had to do more so that readers could react with the same emotion I felt."

BEFORE I HAVE FINISHED WRITING THIS, THE story of how you were born, I will be forty-four years old and the events and feelings which make up the story will be at least eight months old. You are lying in the next room in a cotton jump-suit. You have five teeth. You cannot walk. You do not seem interested in crawling. You are sound asleep.

I have put off writing this so long that, now the time is here, I do not want to write it. I cannot think. Laziness. Wooden shutters over the memory. Nothing comes, no pictures, no feelings, but the architecture of the hospital at Camperdown.

You were born in the King George V Hospital in Missenden Road, Camperdown, a building that won an award for its architecture. It was opened during the Second World War, but its post-Bauhaus modern style has its roots in the time before the First World War, with an optimism about the technological future that we may never have again.

I liked this building. I liked its smooth, rounded, shiny corners. I liked its wide stairs, I liked the huge sash-windows, even the big blue-and-white checked tiles: when I remember this building there is sunshine splashed across those tiles, but there were times when it seemed that other memories might triumph and it would be remembered for the harshness of its neon lights and emptiness of the corridors.

A week before you were born, I sat with your mother in a four-bed ward on the eleventh floor of this building. In this ward she received

PAGE 50

blood transfusions from plum-red plastic bags suspended on stainless steel stands. The blood did not always flow smoothly. bags had to be fiddled with, the stand had to be raised, lowered, its drip-rate increased, decreased, inspected by the sister who __ been a political prisoner in Chile, by the sister from the Solomon Islands, by others I don't remember. The blood entered your mother through a needle in her forearm. When the vein collapsed, a new one had to be found. This was caused by a kind of bruising called "tissuing." We soon knew all about tissuing. It made her arm hurt like hell.

She was bright-eyed and animated as always, but her lips had a slight blue tinge and her skin had a tight, translucent quality.

She was in this room on the west because her blood appeared to be dying. Some thought the blood was killing itself. This is what we all feared, none more than me, for when I heard her blood-count was so low, the first thing I thought (stop that thought, cut it off, bury it) was cancer.

This did not necessarily have a lot to do with Alison, but with me, and how I had grown up, with a mother who was preoccupied with cancer and who, going into surgery for suspected breast cancer, begged the doctor to "cut them both off." When my mother's friend Enid Tanner boasted of her hard stomach muscles, my mother envisaged a growth. When her father complained of a sore elbow, my mother threatened the old man: "All right, we'll take you up to Doctor Campbell and she'll cut it off." When I was ten, my mother's brother got cancer and they cut his leg off right up near the hip and took photographs of him, naked, one-legged, to show other doctors the success of the operation.

When I heard your mother's blood-count was low, I was my mother's son. I thought: cancer.

I remembered what Alison had told me of that great tragedy of her grandparents' life, how their son (her uncle) had leukemia, how her grandfather then bought him the car (a Ford Prefect? a Morris Minor?) he had hitherto refused him, how the dying boy had driven for miles and miles, hours and hours while his cells attacked each other.

I tried to stop this thought, to cut it off. It grew again, like a thistle whose root has not been removed and must grow again, every time, stronger and stronger.

The best hematological unit in Australia was on hand to deal with the problem. They worked in the hospital across the road, the Royal Prince Alfred. They were friendly and efficient. They were not at all like I had imagined big hospital specialists to be. They took blood samples, but the blood did not tell them enough. They returned to take marrow from your mother's bones. They brought a needle with them that would give you the horrors if you could see the size of it.

The doctor's specialty was leukemia, but he said to us: "We don't think it's anything really nasty." Thus "nasty" became a code for cancer.

They diagnosed megnoblastic anemia which, although we did not realize it, is the condition of the blood and not the disease itself.

Walking back through the streets in Shimbashi in Tokyo, your mother once told me that a fortune-teller had told her she would die young. It was for this reason — or so I remembered — that she took such care of her health. At the time she told me this, we had not known each other very long. It was July. We had fallen in love in May. We were still stumbling over each other's feelings in the dark. I took this secret of your mother's lightly, not thinking about the weight it must carry, what it might mean to talk about it. I hurt her; we fought, in the street by the Shimbashi railway station, in a street with shop windows advertising cosmetic surgery, in the Dai-Ichi Hotel in the Ginza district of Tokyo, Japan.

When they took the bone marrow from your mother's spine, I held her hand. The needle had a cruel diameter, was less a needle than an instrument for removing a plug. She was very brave. Her wrists seemed too thin, her skin too white and shiny, her eyes too big and bright. She held my hand because of pain. I held hers because I loved her, because I could not think of living if I did not have her. I thought of what she had told me in Tokyo. I wished there was a God I could pray to.

I flew to Canberra on 7 May 1984. It was my forty-first birthday. I had injured my back and should have been lying flat on a board. I had come from a life with a woman which had reached, for both of us, a state of chronic unhappiness. I will tell you the truth: I was on that airplane to Canberra because I hoped I might fall in love. This made me a dangerous person.

There was a playwrights' conference in Canberra. I hoped there would be a woman there who would love me as I would love her. This was a fantasy I had had before, getting onto airplanes to foreign cities, riding in taxis towards hotels in Melbourne, in Adelaide, in Brisbane. I do not mean that I was thinking about sex, or an affair, but that I was looking for someone to spend my life with. Also — and I swear I have not invented this after the fact — I had a vision of your mother's neck.

I hardly knew her. I met her once at a dinner when I hardly noticed her. I met her a second time when I saw, in a meeting room, the back of her neck. We spoke that time, but I was argumentative and I did not think of her in what I can only call "that way."

And yet as the airplane came down to land in Canberra, I saw your mother's neck, and thought: maybe Alison Summers will be there. She was the dramaturge at the Nimrod Theatre. It was a playwrights' conference. She should be there.

And she was. And we fell in love. And we stayed up till four in the morning every morning talking. And there were other men, everywhere, in love with her. I didn't know about the other men. I knew only that I was in love as I had not been since I was eighteen years old. I wanted to marry Alison Summers, and at the end of the first night we had been out together when I walked her to the door of her room, and we had, for the first time, ever so lightly, kissed on the lips — and also, I must tell you, for it was delectable and wonderful, I kissed your mother on her long, beautiful neck — and when we had kissed and patted the air between us and said "all right" a number of times, and I had walked back to my room where I had, because of my back injury, a thin mattress lying flat on the floor, and when I was in this bed, I said, aloud, to the empty room: "I am going to live with Alison."

And I went to sleep so happy I must have been smiling.

She did not know what I told the room. And it was three or four days before I could see her again, three or four days before we could go out together, spend time alone, and I could tell her what I thought.

I had come to Canberra wanting to fall in love. Now I was in love. Who was I in love with? I hardly knew, and yet I knew exactly. I did not realize how beautiful she was. I found that out later. At the beginning I recognized something more potent than beauty: it was a force, a life, an energy. She had such life in her face, in her eyes — those eyes which you inherited — most of all. It was this I loved, this which I recognized so that I could say — having kissed her so lightly — I will live with Alison. And know that I was right.

It was a conference. We were behaving like men and women do at conferences, having affairs. We would not be so sleazy. After four nights staying up till four a.m. we still had not made love. I would creep back to my room, to my mattress on the floor. We talked about everything. Your mother liked me, but I cannot tell you how long it took her to fall in love with me. But I know we were discussing marriages and babies when we had not even been to bed together. That came early one morning when I returned to her room after three hours' sleep. We had not planned to make love there at the conference but there we were, lying on the bed, kissing, and then we were making love, and you were not conceived then, of course, and yet from that time we never ceased thinking of you and when, later in Sydney, we had to learn to adjust to each other's needs, and when we argued, which we did often then, it was you more than anything that kept us together. We wanted you so badly. We loved you before we saw you. We loved you as we made you, in bed in another room, at Lovett Bay.

When your mother came to the eleventh floor of the King George V Hospital, you were almost ready to be born. Every day the sisters came and smeared jelly on your mother's tight, bulging stomach and then stuck a flat little octopus-type sucker to it and listened to the noises you made.

You sounded like soldiers marching on a bridge.

You sounded like short-wave radio.

You sounded like the inside of the sea.

We did not know if you were a boy or a girl, but we called you Sam anyway. When you kicked or turned we said, "Sam's doing his exercises." We said silly things.

When we heard how low Alison's blood-count was, I phoned

the obstetrician to see if you were OK. She said that as long as the mother's count was above six there was no need to worry.

Your mother's count was 6.2. This was very close. I kept worrying that you had been hurt in some way. I could not share this worry for to share it would only be to make it worse. Also I recognize that I have made a whole career out of making my anxieties get up and walk around, not only in my mind, but in the minds of readers. I went to see a naturopath once. We talked about negative emotions — fear and anger. I said to him, "But I use my anger and my fear." I talked about these emotions as if they were chisels and hammers.

This alarmed him considerably.

Your mother is not like this. When the hematologists saw how she looked, they said: "Our feeling is that you don't have anything nasty." They topped her up with blood until her count was twelve and although they had not located the source of her anemia, they sent her home.

A few days later her count was down to just over six.

It seemed as if there was a silent civil war inside her veins and arteries. The number of casualties was appalling.

I think we both got frightened then. I remember coming home to Louisa Road. I remember worrying that I would cry. I remember embracing your mother — and you too, for you were a great bulge between us. I must not cry. I must support her.

I made a meal. It was salad niçoise. The electric lights, in memory, were all ten watts, sapped by misery. I could barely eat. I think we may have watched a funny film on videotape. We repacked the bag that had been unpacked so short a time before. It now seemed likely that your birth was to be induced. If your mother was sick she could not be properly looked after with you inside her. She would be given one more blood transfusion, and then the induction would begin. And that is how your birthday would be on September thirteenth.

Two nights before your birthday I sat with Alison in the four-bed ward, the one facing east, towards Missenden Road. The curtains were drawn around us. I sat on the bed and held her hand. The blood continued its slow viscous drip from the plum-red bag along the clear plastic tube and into her arm. The obstetrician was with us. She stood at the head of the bed, a kind, intelligent woman in her early thirties. We talked about Alison's blood. We asked her what she thought this mystery could be. Really what we wanted was to be told that everything was OK. There was a look on Alison's face when she asked. I cannot describe it, but it was not a face seeking medical "facts."

The obstetrician went through all the things that were not wrong with your mother's blood. She did not have a vitamin B deficiency. She did not have a folic acid deficiency. There was no iron deficiency. She did not have any of the common (and easily fixable) anemias of pregnancy. So what could it be? we asked, really only wishing to be assured it was nothing "nasty."

"Well," said the obstetrician, "at this stage you cannot rule out cancer."

I watched your mother's face. Nothing in her expression showed what she must feel. There was a slight colouring of her cheeks. She nodded. She asked a question or two. She held my hand, but there was no tight squeezing.

The obstetrician asked Alison if she was going to be "all right." Alison said she would "be all right." But when the obstetrician left she left the curtains drawn.

The obstetrician's statement was not of course categorical and not everyone who has cancer dies, but Alison was, at that instant, confronting the thing that we fear most. When the doctor said those words, it was like a dream or a nightmare. I heard them said. And yet they were not said. They could not be said. And when we hugged each other — when the doctor had gone — we pressed our bodies together as we always had before, and if there were tears on our cheeks, there had been tears on our cheeks before. I kissed your mother's eyes. Her hair was wet with her tears. I smoothed her hair on her forehead. My own eyes were swimming. She said: "All right, how are we going to get through all this?"

Now you know her, you know how much like her that is. She is not going to be a victim of anything.

"We'll decide it's going to be OK," she said, "that's all."

And we dried our eyes.

But that night, when she was alone in her bed, waiting for

the sleeping pill to work, she thought: If I die, I'll at least have made this little baby.

When I left your mother I appeared dry-eyed and positive, but my disguise was a frail shell of a thing and it cracked on the stairs and my grief and rage came spilling out in gulps. The halls of the hospital gleamed with polish and vinyl and fluorescent light. The flower-seller on the ground floor had locked up his shop. The foyer was empty. The whisker-shadowed man in admissions was watching television. In Missenden Road two boys in jeans and sand-shoes conducted separate conversations in separate phone booths. Death was not touching them. They turned their backs to each other. One of them — a red-head with a tattoo on his forearm — laughed.

In Missenden Road there were taxis NOT FOR HIRE speeding towards other destinations.

In Missenden Road the bright white lights above the zebra crossings became a luminous sea inside my eyes. Car lights turned into necklaces and ribbons. I was crying, thinking it is not for me to cry: crying is a poison, a negative force; everything will be all right; but I was weeping as if huge balloons of air had to be released from inside my guts. I walked normally. My grief was invisible. A man rushed past me, carrying roses wrapped in cellophane. I got into my car. The floor was littered with car-park tickets from all the previous days of blood transfusions, tests, test results, admission, etc. I drove out of the car-park. I talked aloud.

I told the night I loved Alison Summers. I love you, I love you, you will not die. There were red lights at the Parramatta Road. I sat there, howling, unroadworthy. I love you.

The day after tomorrow there will be a baby. Will the baby have a mother? What would we do if we knew Alison was dying? What would we do so Sam would know his mother? Would we make a videotape? Would we hire a camera? Would we set it up and act for you? Would we talk to you with smiling faces, showing how we were together, how we loved each other? How could we? How could we think of these things?

I was a prisoner in a nightmare driving down Ross Street in Glebe. I passed the Afrikan restaurant where your mother and I ate after first coming to live in Balmain.

All my life I have waited for this woman. This cannot happen.

I thought: Why would it not happen? Every day people are tortured, killed, bombed. Every day babies starve. Every day there is pain and grief, enough to make you howl to the moon forever. Why should we be exempt, I thought, from the pain of life?

What would I do with a baby? How would I look after it? Day after day, minute after minute, by myself. I would be a sad man, forever, marked by the loss of this woman. I would love the baby. I would care for it. I would see, in its features, every day, the face of the woman I had loved more than any other.

When I think of this time, it seems as if it's two in the morning, but it was not. It was ten o'clock at night. I drove home through a landscape of grotesque imaginings.

The house was empty and echoing.

In the nursery everything was waiting for you, all the things we had got for "the baby." We had read so many books about babies, been to classes where we learned about how babies are born, but we still did not understand the purpose of all the little clothes we had folded in the drawers. We did not know which was a swaddle and which was a sheet. We could not have selected the clothes to dress you in.

I drank coffee. I drank wine. I set out to telephone Kathy Lette, Alison's best friend, so she would have this "news" before she spoke to your mother the next day. I say "set out" because each time I began to dial, I thought: I am not going to do this properly. I hung up. I did deep breathing. I calmed myself. I telephoned. Kim Williams, Kathy's husband, answered and said Kathy was not home yet. I thought: she must know. I told Kim, and as I told him the weeping came with it. I could hear myself. I could imagine Kim listening to me. I would sound frightening, grotesque, and less in control than I was. When I had finished frightening him, I went to bed and slept.

I do not remember the next day, only that we were bright and determined. Kathy hugged Alison and wept. I hugged Kathy and wept. There were isolated incidents. We were "handling it." And, besides, you were coming on the next day. You were life, getting stronger and stronger.

I had practical things to worry about. For instance: the bag. The bag was to hold all the things for the labour ward. There was a list for the contents of the bag and these contents were all purchased and ready, but still I must bring them to the hospital early the next morning. I checked the bag. I placed things where I would not forget them. You wouldn't believe the things we had. We had a cassette-player and a tape with soothing music. We had rosemary and lavender oil so I could massage your mother and relax her between contractions. I had a thermos to fill with blocks of frozen orange juice. There were special cold packs to relieve the pain of a backache labour. There were paper pants — your arrival, after all, was not to happen without a great deal of mess. There were socks, because your mother's feet would almost certainly get very cold. I packed all these things, and there was something in the process of this packing which helped overcome my fears and made me concentrate on you, our little baby, already so loved although we did not know your face, had seen no more of you than the ghostly blue image thrown up by the ultrasound in the midst of whose shifting perspectives we had seen your little hand move. ("He waved to us.")

On the morning of the day of your birth I woke early. It was only just light. I had notes stuck on the fridge and laid out on the table. I made coffee and poured it into a thermos. I made the bagel sandwiches your mother and I had planned months before — my lunch. I filled the bagels with a fiery Polish sausage and cheese and gherkins. For your mother, I filled a spray-bottle with Evian water.

It was a Saturday morning and bright and sunny and I knew you would be born but I did not know what it would be like. I drove along Ross Street in Glebe ignorant of the important things I would know that night. I wore grey stretchy trousers and a black shirt which would later be marked by the white juices of your birth. I was excited, but less than you might imagine. I parked at the hospital as I had parked on all those other occasions. I carried the bags up to the eleventh floor. They were heavy.

Alison was in her bed. She looked calm and beautiful. When we kissed, her lips were soft and tender. She said: "This time tomorrow we'll have a little baby."

In our conversation, we used the diminutive a lot. You were always spoken of as "little," as indeed you must really have been, but we would say "little" hand, "little" feet, "little" baby, and thus evoked all our powerful feelings about you.

This term ("little") is so loaded that writers are wary of using it. It is cute, sentimental, "easy." All of sentient life seems programmed to "little." If you watch grown dogs with a pup, a pup they have never seen, they are immediately patient and gentle, even solicitous, with it. If you watched your mother and father holding up a tiny terry-towelling jump-suit in a department store, you would have seen their faces change as they celebrated your "littleness" while, at the same time, making fun of their own responses — they were aware of acting in a way they would have previously thought of as saccharine.

And yet we were not aware of the torrents of emotion your "littleness" would unleash in us, and by the end of September thirteenth we would think it was nothing other than the meaning of life itself.

When I arrived at the hospital with the heavy bags of cassette-players and rosemary oil, I saw a dark-bearded, neat man in a suit sitting out by the landing. This was the hypnotherapist who had arrived to help you come into the world. He was serious, impatient, eager to start. He wanted to start in the pathology ward, but in the end he helped carry the cassette-player, thermoses, sandwiches, massage oil, sponges, paper pants, apple juice, frozen orange blocks, rolling pin, cold packs, and even water down to the labour ward where — on a stainless steel stand eight feet high — the sisters were already hanging the bag of Oxytocin which would ensure that this day was your birthday.

It was a pretty room, by the taste of the time. As I write it is still that time, and I still think it pretty. All the surfaces were hospital surfaces — easy to clean — laminexes, vinyls, materials with a hard shininess, but with colours that were soft pinks and blues and an effect that was unexpectedly pleasant, even sophisticated.

The bed was one of those complicated stainless steel machines which seems so cold and impersonal until you realize all the clever things it can do. In the wall there were sockets with labels like

"Oxygen." The cupboards were filled with paper-wrapped sterile "objects." There was, in short, a seriousness about the room, and when we plugged in the cassette player we took care to make sure we were not using a socket that might be required for something more important.

The hypnotherapist left me to handle the unpacking of the bags. He explained his business to the obstetrician. She told him that eight hours would be a good, fast labour. The hypnotherapist said he and Alison were aiming for three. I don't know what the doctor thought, but I thought there was not a hope in hell.

When the Oxytocin drip had been put into my darling's arm, when the water-clear hormone was entering her veins, one drip every ten seconds (you could hear the machine click when a drip was released), when these pure chemical messages were being delivered to her body, the hypnotherapist attempted to send other messages of a less easily assayable quality.

I tell you the truth: I did not care for this hypnotherapist, this pushy, over-eager fellow taking up all the room in the labour ward. He sat on the right side of the bed. I sat on the left. He made me feel useless. He said: "You are going to have a good labour, a fast labour, a fast labour like the one you have already visualized." Your mother closed her eyes. She had such large, soft lids, such tender and vulnerable coverings of skin. Inside the pink light of the womb, your eyelids were the same. Did you hear the messages your mother was sending to her body and to you? The hypnotherapist said: "After just three hours you are going to deliver a baby, a good, strong, healthy baby. It will be an easy birth, an effortless birth. It will last three hours and you will not tear." On the door the sisters had tacked a sign reading: QUIET PLEASE. HYPNOTHERAPY IN PROGRES. "You are going to be so relaxed, and in a moment you are going to be even more relaxed, more relaxed than you ever have been before. You are feeling yourself going deeper and deeper and when you come to you will be in a state of waking hypnosis and you will respond to the trigger-words Peter will give you during your labour, words which will make you, once again, so relaxed."

My trigger-words were to be "Breathe" and "Relax."

The hypnotherapist gave me his phone number and asked me to call when you were born. But for the moment you had not felt the effects of the Oxytocin on your world and you could not yet have suspected the adventures the day would have in store for you.

You still sounded like the ocean, like soldiers marching across a bridge, like short-wave radio.

On Tuesday nights through the previous winter we had gone to classes in a building where the lifts were always sticking. We had walked up the stairs to a room where pregnant women and their partners had rehearsed birth with dolls, had watched hours of videotapes of exhausted women in labour. We had practised all the different sorts of breathing. We had learned of the different positions for giving birth: the squat, the supported squat, the squat supported by a seated partner. We knew the positions for the first and second stage, for a backache labour, and so on, and so on. We learned birth was a complicated, exhausting, and difficult process. We worried we would forget how to breathe. And yet now the time was here we both felt confident, even though nothing would be like it had been in the birth classes. Your mother was connected to the Oxytocin drip which meant she could not get up and walk around. It meant it was difficult for her to "belly dance" or do most of the things we had spent so many evenings learning about.

In the classes they tell you that the contractions will start far apart, that you should go to the hospital only when they are ten minutes apart: short bursts of pain, but long rests in between. During this period your mother could expect to walk around, to listen to music, to enjoy a massage. However, your birth was not to be like this. This was not because of you. It was because of the Oxytocin. It had a fast, intense effect, like a double Scotch when you're expecting a beer. There were not to be any ten-minute rests, and from the time the labour started it was, almost immediately, fast and furious, with a one-minute contraction followed by more than two minutes of rest.

If there had been time to be frightened, I think I would have been frightened. Your mother was in the grip of pains she could not escape from. She squatted on a bean bag. It was as if her insides were all tangled, and tugged in a battle to the death. Blood ran from her. Fluid like egg-white. I did not know what anything was. I was a man who had wandered onto a battlefield. The blood was bright with oxygen. I wiped your mother's brow. She panted. *Huh-huh-huh-huh.* I ministered to her with sponge and water. I could not take her pain for her. I could do nothing but measure the duration of pain. I had a little red stop-watch you will one day find abandoned in a dusty drawer. (Later your mother asked me what I had felt during labour. I thought only: I must count the seconds of the contraction; I must help Alison breathe, now, now, now; I must get that sponge — there is time to make the water in the sponge cool — now I can remove that bowl and cover it. Perhaps I can reach the bottle of Evian water. God, I'm so thirsty. What did I think during labour? I thought: When this contraction is over I will get to that Evian bottle.)

Somewhere in the middle of this, in these three hours in this room whose only view was a blank screen of frosted glass, I helped your mother climb onto the bed. She was on all fours. In this position she could reach the gas mask. It was nitrous oxide, laughing gas. It did not stop the pain, but it made it less important. For the gas to work your mother had to anticipate the contraction, breathing in gas before it arrived. The sister came and showed me how I could feel the contraction coming with my hand. But I couldn't. We used the stop-watch, but the contractions were not regularly spaced, and sometimes we anticipated them and sometimes not. When we did not get it right, your mother took the full brunt of the pain. She had her face close to the mattress. I sat on the chair beside. My face was close to hers. I held the watch where she could see it. I held her wrist. I can still see the red of her face, the wideness of her eyes as they bulged at the enormous size of the pains that wracked her.

Sisters came and went. They had to see how wide the cervix was. At first it was only two centimetres, not nearly enough room for you to come out. An hour later they announced it was four centimetres. It had to get to nine centimetres before we could even think of you being born. There had to be room for your head (which we had been told was big — well, we were told wrong, weren't we?) and your shoulders to slip through. It felt to your mother that this labour would go on for eight or twelve or twenty hours. That she should endure this intensity of pain for this time was unthinkable. It was like running a hundred-metre race which was stretching to ten miles. She wanted an epidural — a pain blocker.

But when the sister heard this she said: "Oh do try to hang on. You're doing so well."

I went to the sister, like a shop steward.

I said: "My wife wants an epidural, so can you please arrange it?"

The sister agreed to fetch the anesthetist, but there was between us — I admit it now — a silent conspiracy: for although I had pressed the point and she had agreed it was your mother's right, we both believed (I, for my part, on her advice) that if your mother could endure a little longer she could have the birth she wanted — without an epidural.

The anesthetist came and went. The pain was at its worst. A midwife came and inspected your mother. She said: "Ten centimetres."

She said: "Your baby is about to be born."

We kissed, your mother and I. We kissed with soft, passionate lips as we did the day we lay on a bed at Lovett Bay and conceived you. That day the grass outside the window was a brilliant green beneath the vibrant petals of fallen jacaranda.

Outside the penumbra of our consciousness trolleys were wheeled. Sterile bags were cut open. The contractions did not stop, of course.

The obstetrician had not arrived. She was in a car, driving fast towards the hospital.

I heard a midwife say: "Who can deliver in this position?" (It was still unusual, as I learned at that instant, for women to deliver their babies on all fours.)

Someone left the room. Someone entered. Your mother was pressing the gas mask so hard against her face it was making deep indentations on her skin. Her eyes bulged huge.

Someone said: "Well get her, otherwise I'll have to deliver it myself."

The door opened. Bushfire came in.

Bushfire was aboriginal. She was about fifty years old. She was compact and taciturn like a farmer. She had a face that folded in on itself and let out its feelings slowly, selectively. It was a face to trust, and trust especially at this moment when I looked up to see

Bushfire coming through the door in a green gown. She came in a rush, her hands out to have gloves put on.

There was another contraction. I heard the latex snap around Bushfire's wrists. She said: "There it is. I can see your baby's head." It was you. The tip of you, the top of you. You were a new country, a planet, a star seen for the first time. I was not looking at Bushfire. I was looking at your mother. She was alight with love and pain.

"Push," said Bushfire.

Your mother pushed. It was you she was pushing, you that put that look of luminous love on her face, you that made the veins on her forehead bulge and her skin go red.

Then — it seems such a short time later — Bushfire said: "Your baby's head is born."

And then, so quickly in retrospect, but one can no more recall it accurately than one can recall exactly how one made love on a bed when the jacaranda petals were lying like jewels on the grass outside. Soon. Soon we heard you. Soon you slipped out of your mother. Soon you came slithering out not having hurt her, not even having grazed her. You slipped out, as slippery as a little fish, and we heard you cry. Your cry was so much lighter and thinner than I might have expected. I do not mean that it was weak or frail, but that your first cry had a timbre unlike anything I had expected. The joy we felt. Your mother and I kissed again, at that moment.

"My little baby," she said. We were crying with happiness. "My little baby."

I turned to look. I saw you. Skin. Blue-white, shiny-wet.

I said: "It's a boy."

"Look at me," your mother said, meaning: stay with me, be with me, the pain is not over yet, do not leave me now. I turned to her. I kissed her. I was crying, just crying with happiness that you were there.

The room you were born in was quiet, not full of noise and clattering. This is how we wanted it for you. So you could come into the world gently and that you should — as you were now — be put onto your mother's stomach. They wrapped you up. I said: "Couldn't he feel his mother's skin?" They unwrapped you so you could have your skin against hers.

And there you were. It was you. You had a face, the face we had never known. You were so calm. You did not cry or fret. You had big eyes like your mother's. And yet when I looked at you first I saw not your mother and me, but your two grandfathers, your mother's father, my father; and as my father, whom I loved a great deal, had died the year before, I was moved to see that here, in you, he was alive.

Look at the photographs in the album that we took at this time. Look at your mother and how alive she is, how clear her eyes are, how all the red pain has just slipped off her face and left the unmistakable visage of a young woman in love.

We bathed you (I don't know whether this was before or after) in warm water and you accepted this gravely, swimming instinctively.

I held you (I think this must be before), and you were warm and slippery. You had not been bathed when I held you. The obstetrician gave you to me so she could examine your mother. She said: "Here."

I held you against me. I knew then that your mother would not die. I thought: "It's fine, it's all right." I held you against my breast. You smelled of love-making.

A *woman who plays*
TRUMPET
is deported

MICHELLE CLIFF

Michelle Cliff was born in Kingston, Jamaica in 1946. She confesses that identifying the nationality of her country of invention is "a complicated business": "It is Jamaica that forms my writing for the most part...{At the same time} I often feel what Derek Walcott expresses in his poem "The Schooner Flight": 'I had no nation now but the imagination.'"

THIS STORY IS DEDICATED TO THE MEMORY OF VALAIDA SNOW, *trumpet-player, who was liberated — or escaped — from a concentration camp. She weighed sixty-five pounds. She died in 1956 of a cerebral hemorrhage. This was inspired by her story, but it is an imagining.*

She came to me in a dream and said: "Girl, you have no idea how tough it was. I remember once Billie Holiday was lying in a field of clover. Just resting. And a breeze came and the pollen from the clover blew all over her and the police came out of nowhere and arrested her for possession."

"And the stuff was *red* ... it wasn't even white."

A woman. A black woman. A black woman musician. A black woman musician who plays trumpet. A bitch who blows. A lady trumpet-player. A woman with chops.

It is the thirties. She has been fairly successful. For a woman, black, with an instrument not made of her. Not made of flesh but of metal.

Her father told her he could not afford two instruments for his two children and so she would have to learn her brother's horn.

This woman tucks her horn under her arm and packs a satchel and sets her course. Paris first.

This woman flees to Europe. No, *flee* is not the word. Escape? Not quite right.

She wants to be let alone. She wants them to stop asking for vocals in the middle of a riff. She wants them to stop calling her *novelty, wonder*, chasing after her orchid-coloured Mercedes looking for a lift. When her husband gets up to go, she tosses him the keys, tells him to have it washed every now and then, the brass eyeballs polished every now and then — reminds him it's unpaid for and wasn't her idea anyway.

She wants a place to practise her horn, to blow. To blow rings around herself. So she blows the USA and heads out. On a ship.

And this is not one of those I'm travellin'-light-because-my-man-has-gone situations, no, that mess ended a long time before. He belongs in an orchid-coloured Mercedes — although he'll probably paint the damn thing grey. It doesn't do for a man to flaunt, he would say, all the while choosing her dresses and fox furs and cocktail rings.

He belongs back there; she doesn't.

The ship is French. Families abound. The breeze from the ocean rosying childish cheeks, as uniformed women stand by, holding shuttlecocks, storybooks, bottles. Women wrapped in tricolour robes sip bouillion. Men slap cues on the shuffle-board court, disks skimming the polished deck. Where — and this is a claim to fame — Josephine Baker once walked her ocelot or leopard or cheetah.

A state of well-being describes these people, everyone is groomed, clean, fed. She is not interested in them, but glad of the calm they convey. She is not interested in looking into their staterooms, or their lives, to hear the sharp word, the slap of a hand across a girl's mouth, the moans of intimacy.

The ship is French. The steward assigned to her, Senegalese.

They seek each other out by night, after the families have retired. They meet in the covered lifeboats. They communicate through her horn and by his silver drum.

He noticed the horn when he came the very first night at sea to turn down her bed. Pointed at it, her. The next morning introduced her to his drum.

The horn is brass. The drum, silver. Metal beaten into memory, history. She traces her hand along the ridges of silver — horse, spear, warrior. Her finger catches the edge of a breast; lingers. The skin drumhead as tight as anything.

In the covered lifeboats by night they converse, dispersing the silence of the deck, charging the air, upsetting the complacency, the well-being that hovers, to return the next day.

Think of this as a reverse middle passage.

Who is to say he is not her people?

Landfall.

She plays in a club in the Quartier Latin. This is not as simple as it sounds. She got to the club through a man who used to wash dishes beside Langston Hughes at Le Grand Duc who knew a woman back then who did well who is close to Bricktop who knows the owner of the club. The trumpet player met the man who used to wash dishes who now waits tables at another club. They talked and he said, "I know this woman who may be able to help you." Maybe it was simple, lucky. Anyway, the trumpet player negotiated the chain of acquaintances with grace; got the gig.

The air of the club is blue with smoke. Noise. Voices. Glasses do clink. Matches and lighters flare. The pure green of absinthe grows cloudy as water is added from a yellow ceramic pitcher.

So be it.

She lives in a hotel around the corner from the club, on the rue de l'Université. There's not much to the room: table, chair, bed, wardrobe, sink. She doesn't spend much time there. She has movement. She walks the length and breadth of the city. Her pumps crunch against the gravel paths in the parks. Her heels click along the edge of the river. All the time her mind is on her music. She is let alone.

She takes her meals at a restaurant called Polidor. Her food is set on a white paper-covered table. The lights are bright. She sits at the side of a glass-fronted room, makes friends with a waitress and practises her French. *Friends* is too strong; they talk. Her horn is swaddled in purple velvet and rests on a chair next to her, next to the wall. Safe.

Of course, people stare occasionally, those to whom she is unfamiliar. Once in a while someone puts a hand to a mouth to whisper to a companion. Okay. No one said these people were perfect. She is tired — too tired — of seeing the gape-mouthed darky advertising *Le Joyeux Nègre*. Okay? Looming over a square by the Panthéon in all his happy-go-luckiness.

While nearby a Martiniquan hawked *L'Étudiant Noir*.

Joueux Négritude.

A child points to the top of his *crème brulée* and then at her, smiles. Okay.

But no one calls her nigger. Or asks her to leave. Or asks her to sit away from the window at a darker table in the back by the kitchen, hustling her so each course tumbles into another. *Crudités* into *timbale* into *caramel.*

This place suits her fine.

The piano player longs for a baby part-*Africaine.* She says no. Okay.

They pay her to play. She stays in their hotel. Eats their food in a clean, well-lighted place. Pisses in their toilet.

No strange fruit hanging in the Tuileries.

She lives like this for a while, getting news from home from folks who pass through. Asking, "When you coming back?"

"Man, no need for that."

Noting that America is still TOBA (tough on black asses), lady trumpet players still encouraged to vocalize, she remains. She rents a small apartment on Montparnasse, gets a cat, gives her a name, pays an Algerian woman to keep house.

All is well. For a while.

1940. The club in the Quartier Latin is shut tight. Doors boarded. The poster with her face and horn torn across. No word. No word at all. Just murmurs.

The owner has left the city on a freight. He is not riding the rails. Is not being chased by bad debts. He is standing next to his wife, her mother, their children, next to other women, their husbands, men, their wives, children, mothers-in-law, fathers, fathers-in-law, mothers, friends.

The club is shut. This is what she knows. But rumours and murmurs abound.

The piano player drops by the hotel, leaves a note. She leaves Paris. She heads north.

She gets a gig in Copenhagen, standing in for a sister moving out — simple, lucky — again. Safe. Everyone wore the yellow star there — for a time.

1942. She is walking down a street in Copenhagen. The army of occupation picks her up. Not the whole army — just a couple of kids with machine guns.

So this is how it's done.

She found herself in a line of women. And girls. And little children.

The women spoke in languages she did not understand. Spoke them quietly. From the tone she knew they were encouraging their children. She knows — she who has studied the nuance of sound.

Her horn tucked tight under her armpit. Her only baggage.

The women and girls and little children in front of her and behind her wore layers of clothing. It was a warm day. In places seams clanked. They carried what they could on their persons.

Not all spoke. Some were absolutely silent. Eyes moved into this strange place.

Do you know the work of Beethoven?

She has reached the head of the line and is being addressed by a young man in English. She cannot concentrate. She sweats through the velvet wrapped around her horn. All around her women and girls and little children — from which she is apart, yet of — are being taken in three different directions. And this extraordinary question.

A portrait on a schoolteacher's wall. Of a wiry-haired, beetle-browed man. And he was a coloured genius, the teacher told them, and the children shifted in their seats.

Telemann? He wrote some fine pieces for the horn.

The boy has detected the shape of the thing under her arm.

She stares and does not respond. How can she?

The voices of women and girls and little children pierce the summer air as if the sound was being wrenched from their bodies. The sun is bright. Beads of sweat gather at the neck of the young man's tunic.

It should not be hot. It should be drear. Drizzle. Chill. But she knows better. The sun stays bright.

In the distance is a mountain of glass. The light grazes the surface and prisms split into colour.

Midden. A word comes to her. The heaps of shells, bones and teeth. Refuse of the Indians. The mound-builders. That place by the river just outside of town — filled with mystery and childhood imaginings.

A midden builds on the boy's table, as women and girls and little children deposit their valuables.

In the distance another midden builds.

Fool of a girl, she told herself. To have thought she had seen it all. Left it — the worst piece of it — behind her. The body burning — ignited by the tar. The laughter and the fire. And her inheriting the horn.

Surface

TEXTURES

ANITA DESAI

IT WAS HER OWN FAULT, SHE LATER KNEW — BUT how could she have helped it? When she stood, puckering her lips, before the fruit barrow in the market and, after sullen consideration, at last plucked a rather small but nicely ripened melon out of a heap on display, her only thought had been, Is it worth a *rupee* and fifty *paise*? The lichees looked more poetic, in large clusters like some prickly grapes of a charming rose colour, their long stalks and stiff grey leaves tied in a bunch above them — but were expensive. Mangoes were what the children were eagerly waiting for — the boys, she knew, were raiding

Anita Desai was born in Mussoorie, India in 1937. Her father was Bengali; her mother German. She has described herself as a "Lese Ratte," a reading rat — her family's title for a bookworm — and says that by nine years of age, "I was perfectly aware of the power of words over that of mere sensations."

the mango trees in the school compound daily and their stomach-aches were a result, she told them, of the unripe mangoes they ate and for which they carried paper packets of salt to school in their pockets instead of handkerchiefs — but, leave alone the expense, the ones the fruiterer held up to her enticingly were bound to be sharp and sour for all their parakeet shades of rose and saffron; it was still too early for mangoes. So she put the melon in her string bag, rather angrily — paid the man his one *rupee* and fifty *paise* which altered his expression from one of promise and enticement to that of disappointment and contempt, and trailed off towards the vegetable barrow.

That, she later saw, was the beginning of it all, for if the melon seemed puny to her and boring to the children, from the start her husband regarded it with eyes that seemed newly opened. One would have thought he had never seen a melon before. All through the meal his eyes remained fixed on the plate in the centre of the table with its big button of a yellow melon. He left most of his rice and pulses on his plate, to her indignation. While she scolded, he reached out to touch the melon that so captivated him. With one finger he stroked the coarse grain of its rind, rough with the upraised criss-cross of pale veins. Then he ran his fingers up and down the green streaks that divided it into even quarters as by green silk threads, so tenderly. She was clearing away the plates and did not notice till she came back from the kitchen.

"Aren't you going to cut it for us?" she asked, pushing the knife across to him.

He gave her a reproachful look as he picked up the knife and went about dividing the melon into quarter-moon portions with sighs that showed how it pained him.

"Come on, come on," she said, roughly, "the boys have to get back to school."

He handed them their portions and watched them scoop out the icy orange flesh with a fearful expression on his face — as though he were observing cannibals at a feast. She had not the time to pay any attention to it then but later described it as horror. And he did not eat his own slice. When the boys rushed away, he bowed his head over his plate and regarded it.

"Are you going to fall asleep?" she cried, a little frightened.

"Oh no," he said, in that low mumble that always exasperated her — it seemed a sign to her of evasiveness and pusillanimity, this mumble — "Oh no, no." Yet he did not object when she seized the plate and carried it off to the kitchen, merely picked up the knife that was left behind and, picking a flat melon seed off its edge where it had remained stuck, he held it between two fingers, fondling it delicately. Continuing to do this, he left the house.

The melon might have been the apple of knowledge for Harish — so deadly its poison that he did not even need to bite into it to imbibe it: that long, devoted look had been enough. As he walked back to his office which issued ration cards to the population of their town, he looked about him vaguely but with hunger, his eyes resting not on the things on which people's eyes normally rest — signboards, the traffic, the number of an approaching bus — but on such things, normally considered nondescript and unimportant, as the paving stones on which their feet momentarily pressed, the length of wire in a railing at the side of the road, a pattern of grime on the windowpane of a disused printing press... Amongst such things his eyes roved and hunted and, when he was seated at his desk in the office, his eyes continued to slide about — that was Sheila's phrase later: "slide about" — in a musing, calculating way, over the surface of the crowded desk, about the corners of the room, even across the ceiling. He seemed unable to focus them on a file or a card long enough to put to them his signature — they lay unsigned and the people in the queue outside went for another day without rice and sugar and kerosene for their lamps and Janta cookers. Harish searched — slid about, hunted, gazed — and at last found sufficiently interesting a thick book of rules that lay beneath a stack of files. Then his hand reached out — not to pull the book to him or open it, but to run the ball of his thumb across the edges of the pages. In their large number and irregular cut, so closely laid out like some crisp palimpsest, his eyes seemed to find something of riveting interest and his thumb of tactile wonder. All afternoon he massaged the cut edges of the book's seven hundred odd pages, tenderly, wonderingly. All afternoon his eyes gazed upon them with strange devotion. At five o'clock, punctually, the office shut and the queue disintegrated into vociferous grumbles and threats as people went home instead of to the ration shops, empty-handed instead of loaded with those necessary but, to Harish, so dull comestibles.

Although Government service is as hard to depart from as to enter — so many letters to be written, forms to be filled, files to be circulated, petitions to be made that it hardly seems worthwhile — Harish was, after some time, dismissed — time he happily spent judging the difference between white blotting paper and pink (pink is flatter, denser, white spongier) and the texture of blotting paper stained with ink and that which is fresh, that which has been put to melt in a saucer of cold tea and that which has been doused in a pot of ink. Harish was dismissed.

The first few days Sheila stormed and screamed like some shrill, wet hurricane about the house. "How am I to go to market and buy vegetables for dinner? I don't even have enough for that. What am I to feed the boys tonight? No more milk for them. The washerwoman is asking for her bill to be paid. Do you hear? Do you hear? And we shall have to leave this flat. Where shall we go?" He listened — or didn't — sitting on a cushion before her mirror, fingering the small silver box in which she kept the red *kum-kum* that daily cut a gash from one end of her scalp to the other after her toilet. It was of dark, almost blackened silver, with a whole forest embossed on it — banana groves, elephants, peacocks, and jackals. He rubbed his thumb over its cold, raised surface.

After that, she wept. She lay on her bed in a bath of tears and perspiration, and it was only because of the kindness of their neighbours that they did not starve to death the very first week, for even those who most disliked and distrusted Harish — "Always said he looks like a hungry hyena," said Mr. Bhatia who lived below their flat, "not human at all, but like a hungry, hunchbacked hyena hunting along the road" — felt for the distraught wife and the hungry children (who did not really mind as long as there were sour green mangoes to steal and devour) and looked to them. Such delicacies as Harish's family had never known before arrived in stainless steel and brass dishes, with delicate unobtrusiveness. For a while wife and children gorged on sweetmeats made with fresh buffalo milk, on pulses cooked according to grandmother's recipes, on stuffed bread and the first pomegranates of the season. But, although delicious, these offerings came in small quantities and irregularly and soon they were really starving.

"I suppose you want me to take the boys home to my parents," said Sheila bitterly, getting up from the bed. "Any other man would regard that as the worst disgrace of all — but not you. What is my shame to you? I will have to hang my head and crawl home and beg my father to look after us since you won't," and that was what she did. He was sorry, very sorry to see her pack the little silver *kum-kum* box in her black trunk and carry it away.

Soon after, officials of the Ministry of Works, Housing and Land Development came and turned Harish out, cleaned and painted the flat and let in the new tenants who could hardly believe their luck — they had been told so often they couldn't expect a flat in that locality for at least another two years.

The neighbours lost sight of Harish. Once some children reported they had seen him lying under the *pipal* tree at the corner of their school compound, staring fixedly at the red gashes cut into the papery bark and, later, a boy who commuted to school on a suburban train claimed to have seen him on the railway platform, sitting against a railing like some tattered beggar, staring across the criss-cross of shining rails. But next day, when the boy got off the train, he did not see Harish again.

Harish had gone hunting. His slow, silent walk gave him the appearance of sliding rather than walking over the surface of the roads and fields, rather like a snail except that his movement was not as smooth as a snail's but stumbling as if he had only recently become one and was still unused to the pace. Not only his eyes and his hands but even his barefeet seemed to be feeling the earth carefully, in search of an interesting surface. Once he found it, he would pause, his whole body would gently collapse across it and hours — perhaps days — would be devoted to its investigation and worship. Outside the town the land was rocky and bare and this was Harish's especial paradise, each rock having a surface of such exquisite roughness, of such perfection in shape and design, as to keep him occupied and ecstatic for weeks together. Then the river beyond the rock quarries drew him away and there he discovered the joy of fingering silk-smooth stalks and reeds, stems and leaves.

Shepherd children, seeing him stumble about the reeds, plunging thigh-deep into the water in order to pull out a water lily with its cool, sinuous stem, fled screaming, not certain whether this was a man or a hairy water snake. Their mothers came, some with stones and some with canes at the ready, but when they saw Harish, his skin parched to a violet shade, sitting on the bank and gazing at the transparent stem of the lotus, they fell back, crying "Wah!" gathered closer together, advanced, dropped their canes and stones, held their children still by their hair and shoulder, and came to bow to him. Then they hurried back to the village, chattering. They had never had a Swami to themselves, in these arid parts. Nor had they seen a Swami who looked holier, more inhuman than Harish with his matted hair, his blue, starved skin and single-focused eyes. So, in the evening, one brought him a brass vessel of milk, another a little rice. They pushed their children before them and made them drop flowers at his feet. When Harish stooped and felt among the offerings for something his fingers could respond to, they were pleased, they felt accepted. "Swamiji," they whispered, "speak."

Harish did not speak and his silence made him still holier, safer. So they worshipped him, fed and watched over him, interpreting his moves in their own fashion, and Harish, in turn, watched over their offerings and worshipped.

LIVING
like weasels

ANNIE DILLARD

A WEASEL IS WILD. WHO KNOWS WHAT HE thinks? He sleeps in his underground den, his tail draped over his nose. Sometimes he lives in his den for two days without leaving. Outside, he stalks rabbits, mice, muskrats, and birds, killing more bodies than he can eat warm, and often dragging the carcasses home. Obedient to instinct, he bites his prey at the neck, either splitting the jugular vein at the throat or crunching the brain at the base of the skull, and he does not let go. One naturalist refused to kill a weasel who was socketed into his hand deeply as a rattlesnake. The man could in no way pry the tiny weasel off, and he had to walk half a mile to water, the weasel dangling from his palm, and soak him off like a stubborn label.

Annie Dillard was born in Philadelphia in 1945. Of the writer's urgency, she observes, "Every morning you climb several flights of stairs, enter your study, open the French doors, and slide your desk and chair out into the middle of the air....Your work is to keep cranking the flywheel that turns the gears that spin the bolt in the engine of belief that keeps you and your desk in midair." Of the reader's urgency, she asks "Why are we reading, if not in hope of beauty laid bare, life heightened and its deepest mystery probed?"

And once, says Ernest Thompson Seton — once, a man shot an eagle out of the sky. He examined the eagle and found the dry skull of a weasel fixed by the jaws to his throat. The supposition is that the eagle had pounced on the weasel and the weasel swivelled and bit as instinct taught him, tooth to neck, and nearly won. I would like to have seen that eagle from the air a few weeks or months before he was shot: was the whole weasel still attached to his feathered throat, a fur pendant? Or did the eagle eat what he could reach, gutting the living weasel with his talons before his breast, bending his beak, cleaning the beautiful airborne bones?

I have been reading about weasels because I saw one last week. I startled a weasel who startled me, and we exchanged a long glance.

Twenty minutes from my house, through the woods by the quarry and across the highway, is Hollins Pond, a remarkable piece of shallowness where I like to go at sunset and sit on a tree trunk. Hollins Pond is also called Murray's Pond; it covers two acres of bottomland near Tinker Creek with six inches of water and six thousand lily pads. In winter, brown-and-white steers stand in the middle of it, merely dampening their hooves; from the distant shore they look like miracle itself, complete with miracle's nonchalance. Now, in the summer, the steers are gone. The water lilies have blossomed and spread to a green horizontal plane that is terra firma to plodding blackbirds, and tremulous ceiling to black leeches, crayfish, and carp.

This is, mind you, suburbia. It is a five-minute walk in three directions to rows of houses, though none is visible here. There's a 55 mph highway at one end of the pond, and a nesting pair of wood ducks at the other. Under every bush is a muskrat hole or a beer can. The far end is an alternating series of fields and woods, fields and woods, threaded everywhere with motorcycle tracks — in whose bare clay wild turtles lay eggs.

So. I had crossed the highway, stepped over two low barbed-wire fences, and traced the motorcycle path in all gratitude through the wild rose and poison ivy of the pond's shoreline up into high grassy fields. Then I cut down through the woods to the mossy fallen tree where I sit. This tree is excellent. It makes a dry, upholstered bench at the upper, marshy end of the pond, a plush jetty raised from the thorny shore between a shallow blue body of water and a deep blue body of sky.

The sun had just set. I was relaxed on the tree trunk, ensconced in the lap of lichen, watching the lily pads at my feet tremble and part dreamily over the thrusting path of a carp. A yellow bird appeared to my right and flew behind me. It caught my eye; I swivelled around — and the next instant, inexplicably, I was looking down at a weasel, who was looking up at me.

Weasel! I'd never seen one wild before. He was ten inches long, thin as a curve, a muscled ribbon, brown as fruitwood, soft-furred, alert. His face was fierce, small and pointed as a lizard's; he would have made a good arrowhead. There was just a dot of chin, maybe two brown hairs' worth, and then the pure white fur began that spread down his underside. He had two black eyes I didn't see, any more than you see a window.

The weasel was stunned into stillness as he was emerging from beneath an enormous shaggy wild rose bush four feet away. I was stunned into stillness twisted backward on the tree trunk. Our eyes locked, and someone threw away the key.

Our look was as if two lovers, or deadly enemies, met unexpectedly on an overgrown path when each had been thinking of something else: a clearing blow to the gut. It was also a bright blow to the brain, or a sudden beating of brains, with all the charge and intimate grate of rubbed balloons. It emptied our lungs. It felled the forest, moved the fields, and drained the pond; the world dismantled and tumbled into that black hole of eyes. If you and I looked at each other that way, our skulls would split and drop to our shoulders. But we don't. We keep our skulls. So.

He disappeared. This was only last week, and already I don't remember what shattered the enchantment. I think I blinked, I think I retrieved my brain from the weasel's brain, and tried to memorize what I was seeing, and the weasel felt the yank of separation, the careening splashdown into real life and the urgent current of instinct. He vanished under the wild rose. I waited motionless, my mind suddenly full of data and my spirit with pleadings, but he didn't return.

Please do not tell me about "approach-avoidance conflicts." I tell you I've been in that weasel's brain for sixty seconds, and he was in mine. Brains are private places, muttering through unique and secret tapes — but the weasel and I both plugged into another tape simultaneously, for a sweet and shocking time. Can I help it if it was a blank?

What goes on in his brain the rest of the time? What does a weasel think? He won't say. His journal is tracks in clay, a spray of feathers, mouse blood and bone: uncollected, unconnected, loose-leaf, and blown.

I would like to learn, or remember, how to live. I come to Hollins Pond not so much to learn how to live as, frankly, to forget about it. That is, I don't think I can learn from a wild animal how to live in particular — shall I suck warm blood, hold my tail high, walk with my footprints precisely over the prints of my hands? — but I might learn something of mindlessness, something of the purity of living in the physical senses and the dignity of living without bias or motive. The weasel lives in necessity and we live in choice, hating necessity and dying at the last ignobly in its talons. I would like to live as I should, as the weasel lives as he should. And I suspect that for me the way is like the weasel's: open to time and death painlessly, noticing everything, remembering nothing, choosing the given with a fierce and pointed will.

I missed my chance. I should have gone for the throat. I should have lunged for that streak of white under the weasel's chin and held on, held on through mud and into the wild rose, held on for a dearer life. We could live under the wild rose wild as weasels, mute and uncomprehending. I could very calmly go wild. I could live two days in the den, curled, leaning on mouse fur, sniffing bird bones, blinking, licking, breathing musk, my hair tangled in the roots of grasses. Down is a good place to go, where the mind is single. Down is out, out of your ever-loving mind and back to your careless senses. I remember muteness as a prolonged and giddy fast, where every moment is a feast of utterance received. Time and events are merely poured, unremarked, and ingested directly, like blood pulsed into my gut through a jugular vein. Could two live that way? Could two live under the wild rose, and explore by the pond, so that the smooth mind of each is as everywhere present to the other, and as received and as unchallenged, as falling snow?

We could, you know. We can live any way we want. People take vows of poverty, chastity, and obedience — even of silence — by choice. The thing is to stalk your calling in a certain skilled and supple way, to locate the most tender and live spot and plug into that pulse. This is yielding, not fighting. A weasel doesn't "attack" anything; a weasel lives as he's meant to, yielding at every moment to the perfect freedom of single necessity.

I think it would be well, and proper, and obedient, and pure, to grasp your one necessity and not let it go, to dangle from it limp wherever it takes you. Then even death, where you're going no matter how you live, cannot you part. Seize it and let it seize you up aloft even, till your eyes burn out and drop; let your musky flesh fall off in shreds, and let your very bones unhinge and scatter, loosened over fields, over fields and woods, lightly, thoughtless, from any height at all, from as high as eagles.

A
WEDGE
of shade

LOUISE ERDRICH

Louise Erdrich was born in Little Falls, Minnesota in 1954 and grew up in Wakpeton, North Dakota near the Turtle Mountain Reservation of which her Chippewa grandfather was the tribal chairman. She describes the first stage of her writing career: "My father used to give me a nickel for every story I wrote, and my mother wove strips of construction paper together and stapled them into book covers. So at an early age I felt myself to be a published author earning substantial royalties. Mine were wonderful parents; they got me excited about reading and writing in a lasting way."

EVERY PLACE THAT I COULD NAME YOU, IN THE whole world around us, has better things about it than Argus. I just happened to grow up there for eighteen years and the soil got to be part of me, the air has something in it that I breathed. Argus water, fluoridated by an order of the state, doesn't taste as good as water in the cities. Still, the first thing I do, walking back into my mother's house, is stand at the kitchen sink and toss down glass after glass.

"Are you filled up?" My mother stands behind me. "Sit down if you are."

She's tall and broad square, with long arms and big knuckles. Her face is rawboned, fierce, and almost masculine in its edges and planes. Several months ago, a beauty operator convinced her that she should feminize her look with curls. Now the permanent, grown out in grizzled streaks, bristles like the coat of a terrier. I don't look like her. Not just the hair, since hers is salt and pepper, mine is a reddish brown, but my build. I'm short, boxy, more like my Aunt Mary, although there's not much about me that corresponds even to her, except it's true that I can't seem to shake this town. I keep coming back here.

"There's jobs at the beet plant."

This rumour, probably false as the plant is in a slump, drops into the dim close air of the kitchen. We have the shades drawn because it's a hot June, over a hundred degrees, and we're trying to stay cool. Outside, the water has been sucked from everything. The veins in the leaves are hollow, the ditch grass is crackling. The sky has absorbed every drop. It's a thin whitish blue veil stretched from end to end over us, a flat gauze tarp. From the depot, I've walked here beneath it, dragging my suitcase.

We're sweating like we're in an oven, a big messy one. For a week, it's been too hot to move much or even clean, and the crops are stunted, failing. The farmer next to us just sold his field for a subdivision, but the workers aren't doing much. They're wearing wet rags on their heads, sitting near the house sites in the brilliance of noon. The studs of woods stand uselessly upright, over them. Nothing casts a shadow. The sun has dried them up too.

"The beet plant," my mother says again.

"Maybe so," I say, and then, because I've got something bigger on my mind, "maybe I'll go out there and apply."

"Oh?" She is intrigued now.

"God, this is terrible!" I take the glass of water in my hand and tip some on my head. I don't feel cooler though, I just feel the steam rising off me.

"The fan broke down," she states. "Both of them are kaput now. The motors or something. If Mary would get the damn tax refund we'd run out to Pamida, buy a couple more, set up a breeze. Then we'd be cool out here."

"Your garden must be dead," I say, lifting the edge of the pull shade.

"It's sick, but I watered. And I won't mulch, that draws the damn slugs."

"Nothing could live out there, no bug." My eyes smart from even looking at the yard, cleared on the north, almost incandescent.

"You'd be surprised."

I wish I could blurt it out, just tell her. Even now, the words swell in my mouth, the one sentence, but I'm scared and with good reason. There is this about my mother: it is awful to see her angry. Her lips press together and she stiffens herself within, growing wooden, silent. Her features become fixed and remote, she will not speak. It takes a long time, and until she does you are held in suspense. Nothing that she ever says, in the end, is as bad as that feeling of dread. So I wait, half believing that she'll figure out my secret for herself, or drag it out of me, not that she ever tries. If I'm silent, she hardly notices. She's not like Aunt Mary, who forces me to say more than I know is on my mind.

My mother sighs, "It's too hot to bake. It's too hot to cook. But it's too hot to eat, anyway." She's talking to herself, which makes me reckless. Perhaps she is so preoccupied by the heat that I can slip my announcement past her. I should just say it, but I lose nerve, make an introduction that alerts her.

"I have something to tell you."

I've cast my lot, there's no going back unless I think quickly. My thoughts hum.

But she waits, forgetting the heat for a moment.

"Ice," I say, "we have to have ice." I speak intensely, leaning toward her, almost glaring, but she is not fooled.

"Don't make me laugh," she says, "there's not a cube in town. The refrigerators can't keep cold enough."

She eyes me as if I'm an animal about to pop from its den and run.

"Okay." I break down. "I really do have something." I stand, turn my back. In this lightless warmth I'm dizzy, almost sick. Now I've gotten to her and she's frightened to hear, breathless.

"Tell me," she urges. "Go on, get it over with."

And so I say it. "I got married." There is a surge of relief, as if a wind blows through the room, but then it's gone. The curtain flaps and we're caught again, stunned in an even denser heat. It's now my turn to wait, and I whirl around and sit right across from her. But I can't bear the picture she makes, the shock that parts her lips, the stunned shade of hurt in her eyes. I have to convince her, somehow, that it's all right.

"You hate weddings! Just think, just picture it. Me, white net. On a day like this. You, stuffed in your summer wool, and Aunt Mary, God knows... and the tux, the rental, the groom..."

Her head lowered as my words fell on her, but now her forehead tips up and her eyes come into view, already hardening. My tongue flies back into my mouth.

She mimics, making it a question, "The groom..."

I'm caught, my lips half open, a stuttering noise in my throat. How to begin? I have rehearsed this but my lines melt away, my opening, my casual introductions. I can think of nothing that would, even in a small way, convey any part of who he is. There is no picture adequate, no representation that captures him. So I just put my hand across the table, and I touch her hand.

"Mother," I say, like we're in a staged drama, "he'll arrive here shortly."

There is something forming in her, some reaction. I am afraid to let it take complete shape.

"Let's go out and wait on the steps, Mom. Then you'll see him."

"I do not understand," she says in a frighteningly neutral voice. This is what I mean. Everything is suddenly forced, unnatural, as though we're reading lines.

"He'll approach from a distance." I can't help speaking like a bad actor. "I told him to give me an hour. He'll wait, then he'll come walking down the road."

We rise and unstick our blouses from our stomachs, our skirts from the backs of our legs. Then we walk out front in single file, me behind, and settle ourselves on the middle step. A scrubby box elder tree on one side casts a light shade, and the dusty lilacs seem to catch a little breeze on the other. It's not so bad out here, still hot, but not so dim, contained. It's worse past the trees. The heat shimmers in a band, rising off the fields, out of the spars and bones of houses that will wreck our view. The horizon and the edge of town show through the spacing now, and as we sit we watch the workers move, slowly, almost in a practised recital, back and forth. Their headcloths hang to their shoulders, their hard hats are dabs of yellow, their white T-shirts blend into the fierce air and sky. They don't seem to be doing anything, although we hear faint thuds from their hammers. Otherwise, except for the whistles of a few birds, there is silence. We certainly don't speak.

It is a longer wait than I anticipated, maybe because he wants to give me time. At last the shadows creep out, hard, hot, charred, and the heat begins to lengthen and settle. We are going into the worst of the afternoon, when a dot at the end of the road begins to form.

Mom and I are both watching. We have not moved our eyes around much, and we blink and squint to try and focus. The dot doesn't change, not for a long while. And then it suddenly springs clear in relief, a silhouette, lost a moment in the shimmer, reappearing. In that shining expanse he is a little wedge of moving shade. He continues, growing imperceptibly, until there are variations in the outline, and it can be seen that he is large. As he passes the construction workers, they turn and stop, all alike in their hats, stock-still.

Growing larger yet as if he has absorbed their stares, he nears us. Now we can see the details. He is dark, the first thing. I have not told my mother, but he's Chippewa, from the same tribe as she. His arms are thick, his chest is huge and the features of his face are wide and open. He carries nothing in his hand. He wears a black T-shirt, the opposite of the construction workers, and soft jogging shoes. His jeans are held under his stomach by a belt with a star beaded on the buckle. His hair is long, in a tail. I am the wrong woman for him. I am paler, shorter, unmagnificent. But I stand up. Mom joins me, and I answer proudly when she asks, "His name?"

"His name is Gerry."

We descend one step, and stop again. It is here we will receive him. Our hands are folded at our waists. We're balanced, composed. He continues to stroll toward us, his white smile widening, his eyes filling with the sight of me as mine are filling with him. At the end of the road, behind him, another dot has appeared. It is fast-moving and the sun flares off it twice, a vehicle. Now there are two figures. One approaching in a spume of dust from the rear, and Gerry, unmindful, not slackening or quickening his pace, continuing on. It is like a choreography design. They move at parallel speeds, in front of our eyes. At the same moment, at the end of our yard, as if we have concluded a performance now, both of them halt.

Gerry stands, looking toward us, his thumbs in his belt. He nods respectfully to Mom, looks calmly at me, and half smiles. He raises his brows, and we're suspended. Officer Lovchik emerges from the police car, stooped and tired. He walks up behind Gerry and I hear the snap of handcuffs, then I jump. I'm stopped by Gerry's gaze though, as he backs away from me, still smiling tenderly. I am paralysed halfway down the walk. He kisses the air while Lovchik cautiously prods at him, fitting his prize into the car. And then the doors slam, the engine roars and they back out and turn around. As they move away there is no siren. I think I've heard Lovchik mention questioning. I'm sure it is lots of fuss for nothing, a mistake, but it cannot be denied, this is terrible timing.

I shake my shoulders, smooth my skirt and turn to Mother with a look of outrage.

"How do you like that?" I try.

She's got her purse in one hand, her car keys out.

"Let's go," she says.

"Okay," I answer. "Fine. Where?"

"Aunt Mary's."

"I'd rather go and bail him out, Mom."

"Bail," she says, "*bail?*"

She gives me such a look of cold and furious surprise that I sink immediately into the front seat, lean back against the vinyl. I almost welcome the sting of the heated plastic on my back, thighs, shoulders.

Aunt Mary's dogs are rugs in the dirt, flattened by the heat of the day. Not one of them barks at us to warn her. We step over them and get no more reaction than a whine, the slow beat of a tail. Inside, we get no answers either, although we call Aunt Mary up and down the hall. We enter the kitchen and sit at the table which contains a half-ruined watermelon. By the sink, in a tin box, are cigarettes. My mother takes one and carefully puts the match to it, frowning.

"I know what," she says. "Go check the lockers."

There are two, a big freezer full of labelled meats and rental space, and another, smaller one that is just a side cooler. I notice, walking past the display counter, that the red beacon beside the outside switch of the cooler is glowing. That tells you when the light is on inside.

I pull the long metal handle toward me and the thick door swishes open. I step into the cool, spicy air. She is there, too proud to ever register a hint of surprise. Aunt Mary simply nods and looks away, as though I've just gone out for a minute, although we've not seen one another in six months or more. She is relaxing, reading a scientific magazine article. I sit down on a barrel of alum labelled Zanzibar and drop my bomb with no warning. "I'm married." It doesn't matter how I tell it to Aunt Mary, because she won't be, refuses to be, surprised.

"What's he do?" she simply asks, putting aside the sheaf of paper. I thought the first thing she'd do is scold me for fooling my mother. But it's odd. For two women who have lived through boring times and disasters, how rarely one comes to the other's defence, and how often they are willing to take advantage of the other's absence. But I'm benefitting here. It seems that Aunt Mary is truly interested in Gerry. So I'm honest.

"He's something like a political activist. I mean he's been in jail and all. But not for any crime, you see, it's just because of his convictions."

She gives me a long, shrewd stare. Her skin is too tough to wrinkle, but she doesn't look young. All around us hang loops of sausages, every kind you can imagine, every colour from the purple-black of blutwurst to the pale whitish links that my mother likes best. Blocks of butter and headcheese, a can of raw milk, wrapped parcels and cured bacons are stuffed onto the shelves around us. My heart has gone still and cool inside of me, and I can't stop talking.

"He's the kind of guy it's hard to describe, very different. People call him a free spirit, but that doesn't say it either because he's very disciplined in some ways. He learned to be neat in jail." I pause, she says nothing, so I go on. "I know it's sudden, but who likes wedding? I hate them, all that mess with the bridesmaids' gowns, getting material to match. I don't have girl friends, I mean, how embarrassing, right? Who would sing 'Oh Perfect Love?' Carry the ring?"

She isn't really listening.

"What's he do?" she asks again.

Maybe she won't let go of it until I discover the right answer, like a game with nouns and synonyms.

"He, well he agitates," I tell her.

"Is that some kind of factory work?"

"Not exactly, no, it's not a nine-to-five job or anything..."

She lets the pages fall, now, cocks her head to the side and stares at me without blinking her cold yellow eyes. She has the look of a hawk, of a person who can see into the future but won't tell you about it. She's lost business for staring at customers, but she doesn't care.

"Are you telling me that he doesn't" — here she shakes her head twice, slowly, from one side to the other without removing me from her stare — "that he doesn't have regular work?"

"Oh, what's the matter anyway," I say roughly. "I'll work. This is the nineteen seventies."

She jumps to her feet, stands over me, a stocky woman with terse features and short, thin points of gray hair. Her earrings tremble and flash, small fiery opals. Her brown plastic glasses hang crooked on a cord around her neck. I have never seen her become quite so instantaneously furious, so disturbed.

"We're going to fix that," she says.

The cooler instantly feels smaller, the sausages knock at my shoulder and the harsh light makes me blink. I am as stubborn as Aunt Mary, however, and she knows that I can go head-to-head with her.

"We're married and that's final." I manage to stamp my foot.

Aunt Mary throws an arm back, blows air through her cheeks and waves away my statement vigorously.

"You're a little girl. How old is he?"

I frown at my lap, trace the threads in my blue cotton skirt and tell her that age is irrelevant.

"Big word," she says sarcastically. "Let me ask you this. He's old enough to get a job?"

"Of course he is, what do you think. Okay, he's older than me. He's in his thirties."

"Aha, I knew it."

"Geez! So what? I mean, haven't you ever been in love, hasn't someone ever gotten you *right here*?" I smash my fist on my chest. We lock eyes, but she doesn't waste a second in feeling hurt.

"Sure, sure I've been in love. You think I haven't? I know what it feels like, you smartass. You'd be surprised. But he was no lazy sonofabitch. Now listen..." She stops, draws her breath, and I let her. "Here's what I mean by 'fix.' I'll teach the sausage-making trade to him, you too, and the grocery business. I've about had it anyway, and so's your mother. We'll do the same as my aunt and uncle — leave the shop to you and move to Arizona. I like this place." She looks up at the burning safety bulb, down to me again. Her face drags in the light. "But what the hell. I always wanted to travel."

I'm kind of stunned, a little flattened out, maybe ashamed of myself.

"You hate going anywhere," I say, which is true.

The doors swing open and Mom comes in with us. She finds a can and balances herself, sighing at the delicious feeling of the air, absorbing from the silence the fact we have talked. She hasn't anything to add, I guess, and as the coolness hits her eyes fall shut. Aunt Mary too. I can't help it either, and my eyelids drop although my brain is alert and conscious. From the darkness, I can see us in the brilliance. The light rains down on us. We sit the way we have been sitting, on our cans of milk and flour, upright and still. Our hands are curled loosely in our laps. Our faces are blank as the gods. We could be statues in a tomb sunk into the side of a mountain. We could be dreaming the world up in our brains.

It is later and the weather has no mercy. We are drained of everything but simple thoughts. It's too hot for feelings. Driving home, we see how field after field of beets has gone into shock, and even some of the soybeans. The plants splay, limp, burned into the ground. Only the sunflowers continue to struggle upright, bristling but small.

What drew me in the first place to Gerry was the unexpected. I went to hear him talk just after I enrolled at the U of M and then I demonstrated when they came and got him off the stage. He always went so willingly, accommodating everyone. I began to visit him. I sold lunar calendars and posters to raise his bail and eventually free him. One thing led to another and one night we found ourselves alone in a Howard Johnson's where they put him up when his speech was finished. There were more beautiful women after him, he could have had his pick of Swedes or Yankton Sioux girls, who are the best-looking of all. And then there was no going back once it started, no turning, as though it were meant. We had no choice.

I have this intuition as we near the house, in the fateful quality of light, as in the turn of the day the heat continues to press and the blackness, into which the warmth usually lifts, lowers steadily. We must come to the end of something. There must be a close to this day.

As we turn into the yard we see that Gerry is sitting on the stairs. Now it is our turn to be received. I throw the car door open and stumble out before the motor even cuts. I run to him and hold him, as my mother, pursuing the order of events, parks carefully. Then she walks over too, holding her purse by the strap. She stands before him and says no word but simply looks into his face as if he's cardboard, a man behind glass who cannot see her. I think she's rude, but then I realize that he is staring back, that they are the same height. Their eyes are level. He puts his hand out.

"My name is Gerry."

"Gerry what?"

"Nanapush."

She nods, shifts her weight. "You're from that line, old strain, the ones . . ." She does not finish.

"And my father," Gerry says, "was Old Man Pillager." He has said this before but I never heard any special meaning in it.

"Kashpaws," she says, "are my branch of course. We're probably related through my mother's brother." They do not move. They are like two opponents from the same divided country, staring across the border. They do not shift or blink and I see that they are more alike than I am like either one of them, so tall, solid, dark-haired. She could be the mother, he the son.

"Well, I guess you should come in," she offers, "you are a distant relative after all." She looks at me. "Distant enough."

Whole swarms of mosquitoes are whining down, discovering us now, so there is no question of staying where we are. And so we walk into the house, much hotter than outside with the gathered heat. Instantly the sweat springs from our skin and I can think of nothing else but cooling off. I try to force the windows higher in their sashes, but there's no breeze anyway, nothing stirs, no air.

"Are you sure," I gasp, "about those fans?"

"Oh, they're broke," my mother says, distressed. I rarely hear this in her voice. She switches on the lights, which makes the room seem hotter, and we lower ourselves into the easy chairs. Our words echo as though the walls have baked and dried hollow.

"Show me those fans," says Gerry.

My mother points toward the kitchen. "They're sitting on the table. I've already tinkered with them. See what you can do."

And so he does. After a while she hoists herself and walks out back with him. Their voices close together now, absorbed, and their tools clank frantically as if they are fighting a duel. But it is a race with the bell of darkness and their waning energy. I think of ice. I get ice on the brain.

"Be right back," I call out, taking the keys from my mother's purse, "do you need anything?"

There is no answer from the kitchen but a furious sputter of metal, the clatter of nuts and bolts spilling to the floor.

I drive out to the Super Pumper, a big new gas-station complex on the edge of town where my mother most likely has never been. She doesn't know about convenience stores, has no credit cards for groceries, gas, pays only with small bills and change. She never has used an ice machine. It would grate on her that a bag of frozen water costs eighty cents, but it doesn't bother me. I take the Styrofoam cooler and I fill it for a couple dollars. I buy two six-packs of Shasta sodas and I plunge then into the uniform coins of ice. I drink two myself, on the way home, and I manage to lift the whole heavy cooler out of the trunk, carry it to the door.

The fans are whirling, beating the air.

I hear them going in the living room the minute I come in. The only light shines from the kitchen. Gerry and my mother have thrown pillows from the couch onto the living room floor, and they are sitting in the rippling currents of air. I bring the cooler in and put it near us. I have chosen all dark flavours — black cherry, grape, black raspberry, so as we drink it almost seems the darkness swirls inside us with the night air, sweet and sharp, driven by small motors.

I drag more pillows down from the other rooms upstairs. There is no question of attempting the bedrooms, the stifling beds. And so, in the dark, I hold hands with Gerry as he settles down between my mother and me. He is huge as a hill between the two of us, solid in the beating wind.

Inside
MEMORY

TIMOTHY FINDLEY

*Timothy Findley was born
in Toronto in 1930,
has won the Governor
General's Award, the Trillium
Award, and is an officer of the
Order of Canada.*

Stone Orchard
November 11, 1970
Radio

IN THE PLAYS OF ANTON CHEKHOV, THERE IS always a moment of profound silence, broken by the words: "I remember..." What follows inevitably breaks your heart. A woman will stand there and others will sit and listen and she will say: "I remember the band playing and the firing at the cemetery as they carried the coffin. Though he was a general, in command of a brigade, yet, there weren't many people there. It was raining. Heavy rain and snow."

*He considers, "Some writing
is like talking to yourself.
That is why, so often, I fail
to get my points across.
Like Jo {Diblee} said
about my piano playing:
"It doesn't matter, dear.
I know you hear the rest
of it in your mind."
Perhaps that is what
a first draft is:
the unaccompanied piano —
and the other drafts are the
orchestration. Ultimately, it is
all one concerto — but the
reader plays the solo part —
which, in the beginning,
the writer played."*

Or some such thing. And she is transformed, this woman, by her memories — absolutely transformed. And as you watch her and listen to her, you are transformed, too — or something inside you is. You change. Your attitude changes. In a way — if it has been well done — your life changes. Why should this be?

I think one reason must be that Chekhov discovered the dramatic value of memory — that a woman in tears remembers happiness; that a smiling, laughing man remembers pain. This gives you two views in one: depth and contrast. But, there's more to it than that. Memory, Chekhov also discovered, is the means by which most of us retain our sanity. The act of remembrance is good for people. Cathartic. Memory is the purgative by which we rid ourselves of the present.

Because memory is what it is, the first thing we tend to "remember" is that time passes. In going back, we recognize that we've survived the passage of time — and if we've survived what we remember, then it's likely we'll survive the present. Memory is a form of hope.

If the memory is a bad one, say of pain or of a death — then it's clouded. The sharpness is blunted. We remember that we were in pain. But the pain itself cannot be recalled exactly. Not as it was. Because, if we could recall it, then we'd have to be in pain again — and that, except where there's psychological disorder, is a physical and mental impossibility. If you've ever had a bad accident, then you'll remember that you can't remember what happened. But you can recall joy. You can make yourself laugh again and feel again something joyous that happened before.

Of course, you can make yourself cry again, too. But the tears aren't as valid as the laughter, because the tears you conjure have as much to do with the passage of time as with the sadness you remember. Still, a sad memory is better than none. It reminds you of survival.

Most of the activity in your brain relies on memory. That takes energy. Have you ever noticed that when you're tired and there's silence in your brain, you begin to sing? That's good health taking over. The tensions of serious thought are being released through play.

Today is Remembrance Day, and it's a strange thing to me that we confine ourselves to remembering only the dead — and only the war dead. If they were able, what would *they* be remembering? Us. And we're alive. Here we are. Maybe it's sad — I suppose it is — that the dead should be remembering the living and the living remembering the dead, But the main thing is, we all remember when we were together. We remember what we were in another time. Not now, but *then*. Memory is making peace with time.

They say that loss of memory is not to know who you are. Then, I suppose, it has to follow that we *are* what we remember. I can believe that. I mean, it's very easy for me to imagine forgetting my name. That wouldn't worry me. And it wouldn't worry me to forget how old I am (I wish I could!) or to forget the colour of my eyes and have to go look in a mirror to remind myself. None of that would worry me. Because I can skip all of that. None of those things are who I am.

But it would worry the hell out of me if I couldn't remember the smell of the house where I grew up, or the sound of my father playing the piano, or the tune of his favourite song. I remember my brother, Michael, as a child. And the child I remember being myself is as much a remembrance of him as it is of me. More, in fact — because I saw him every day and did not see myself. I heard him every day — and did not hear myself (except singing). So, to be a child in memory means that I conjure Michael, not the child in photographs who bears my name.

I am my Aunt Marg, for instance, telling me not to lean into the cemetery over the fence at Foxbar Road. I am not me leaning over the fence, I am her voice — because that is what I remember. And I am all the gravestones I was looking at when she called me. And the fence boards that supported me. And the sun on my back. But I am not that little boy. I don't remember him at all. I remember him falling and being picked up — but I am the distance he fell and the hands that lifted him, not the bump in between. I remember the sound of my own voice crying — but not the feel of it. That voice is gone. And I am the gloves my mother wore when she held my hand and the tones of her laughter. And I remember and will move forever, as all children do, to the heartbeats of my mother. That remembrance is the rhythm of my life. So memory is other people — it is little of ourselves.

I like Remembrance Day. I'm fond of memory. I wish it was a day of happiness. I have many dead in my past, but only one of them died from the wars. And I think very fondly of him. He was my uncle. He didn't die in the War, but because of it. This was the First World War and so I don't remember the event itself. I just remember him. But what I remember of my uncle is not the least bit sad.

I was just a child — in the classic sense — a burbling, few-worded, looking-up-at-everything child. Uncle Tif — who died at home — was always in a great tall bed — high up — and the bed was white. I would go into his room supported by my father's hands, and lean against the lower edge of the mattress. There was a white sheet over everything, and I can smell that sheet to this day. It smelled of soap and talcum powder. To me, Uncle Tif was a hand that came down from a great way off and tapped me on the head. He smoked a pipe. And there was something blue in the room — I don't know whether it was a chair or a table or my father's pant legs — but there was something blue and that has always been one of my favourite colours.

And high above my head, there was a tall glass jar on a table and the jar was full of hard French candies. They had shiny jackets and were many colours. And Uncle Tif's hand would go out, waving in the air above my gaze and lift up the lid of the jar and take out a candy and slowly — it was always slowly — he would pass the candy down into my open mouth. Then I would lean against the bed, or fall on the floor, and taste the candy for about two hours — or what, to a kid, just seemed two hours — while the adult voices buzzed above my head.

I know he sacrificed his youth, his health, his leg, and finally his life for his country. But I'd be a fool if I just said *thanks* — *I'm grateful*. I might as well hit him in the mouth as say that. Because my being grateful has nothing to do with what he died for or why he died. That was part of his own life and what I am grateful for is that he had his own life. I am grateful he was there in that little bit of my life. And I am grateful, above all, that he is in my memory, I am his namesake. He is mine.

Remembrance is more than honouring the dead. Remembrance is joining them — being one with them in memory. Memory is survival.

GOTCHA!

ROBERT FULFORD

IN THE COURSE OF A SHAKESPEAREAN production in Toronto in 1987, there was a moment that briefly illustrated why contemporary society desperately needs literature and the literary imagination. The moment came just after the scene in *Henry V* in which some soldiers, about to leave for war, tearfully said good-bye to their wives. As soon as the women were safely out of sight, martial music poured from loudspeakers, the men shouted with joy, and patriotic signs were paraded across the stage. One sign held a single word: "Gotcha!"

Robert Fulford was born in Ottawa in 1932 and is descended from several generations of Canadian newspapermen. He is an officer of the Order of Canada. He has said that the only interesting idea he ever heard in high school was from a business teacher who defined the purpose of economics as activity used to support society so that society could support the arts; "the arts, in turn, made life worth living, thus justifying all the attention we gave to the economy."

What was remarkable about that little piece of modernized Shakespeare was that it placed, in the middle of a work from the greatest literary imagination of the ages, a graphic reminder of the 20th-century imagination at its meanest and most degraded.

Not everyone in the Canadian audience understood why "Gotcha!" was there. This was the English Shakespeare Company, and the reference was to something that happened in England five years earlier. On the afternoon of May 3, 1982, west of the Falkland Islands, torpedos from a British submarine hit the General Belgrano, an Argentine cruiser. Almost immediately the ship began to sink. When the news of this victory reached London, the *Sun*, a hugely successful tabloid, put a one-word headline on the next morning's front page: Gotcha!

This quickly became famous as a symbol of blind jingoism, but it was also a spectacular instance of failed imagination. The people who put that headline on their newspaper were victims of the peculiar callousness that afflicts all of us to some degree. What they did was hideously inappropriate, but it was also in a sense consistent with their training, and consistent with the atmosphere of this period in history.

During the sinking, about 300 sailors, many of them teenage conscripts, choked to death on smoke, burned to death in oil or boiling water, or sank to the bottom of the sea. The rest of the crew, 800 or so, spent 36 hours floating on rafts in icy water, praying for rescue. The appropriate response to any such event is pity and terror, but the response of the people at the *Sun* was boyish glee. The *Sun* had already been treating the Falklands War as a kind of video game, a clash of abstract forces with no human meaning. The ships, the submarines, the helicopters and the people on them were no more consequential than flickers of electric light on a screen.

Flickers of light are the problem — perhaps the greatest mass emotional problem of our era. Flickers of light on the television screen, or the movie screen, have become our principal means of receiving information about distant reality. Television brings us close to certain forms of reality, such as war in the Persian Gulf, but it also separates us emotionally from whatever it shows us. The more we see, the less we feel. Television instructs us that one war looks much like another, one plane crash much like another; we lose our sense of the human meaning of disaster. Mass communication deadens rather than enlivens us.

In the movies, too, we learn that the death of others is unimportant. For a quarter-century the movies have been teaching us that people who die by gunfire are usually only extras, or deserve to die.

Those who defend violence in entertainment are quick to point out that it has always been part of drama and literature — there's violence in the Bible, in the Greek tragedies and, of course, in Shakespeare. But until our time, violence in drama and literature was given meaning. It was given weight. It was set in a context that made the appropriate response — pity and terror — possible. In Shakespeare, no one dies without a purpose. One moral of the Shakespeare history plays is that those who kill their kings will live to rue it. Certainly those plays tell us, again and again, that the results of killing are never negligible — and that they will be felt for generations.

On the other hand, the editor who wrote: "Gotcha!" later said, "I agree that headline was a shame. But it wasn't meant in a blood-curdling way. We just felt excited and euphoric. Only when we began to hear reports of how many men died did we begin to have second thoughts." There speaks a sadly crippled imagination, desperately in need of literature.

The future of literature is in question. The novel is no longer, for most people, the central means of expressing a culture. Poetry is read by only a few. Literary studies no longer stand at the centre of the university curriculum. Some of literature's tasks, such as social observation, are often accomplished better by movies and TV programs. Even in the bookstores, literature is often pushed aside by journalism, how-to manuals and cookbooks.

But literature remains the core of civilized life precisely because it is the only reliable antidote to everything in our existence that diminishes us. Only the literary imagination can save us from the deadening influence of visual news and visual entertainment. When it works as it should, literature takes us beyond our parochialism into other minds and other cultures. It makes us know that even our enemies, even anonymous Argentinian sailors, are as humanly diverse as we are.

If we let it, literature can also save us from the narrowing effect of politics. Politics teaches us to see the world in functional terms, defined by power blocs and national borders and pressure groups. Pretending to offer freedom, politics asks us to identify ourselves by ethnicity or gender or class or nationality. Literature, on the other hand, dares us to feel our way across all boundaries of thought and feeling.

One of the more beautiful stories I've read in recent years was written by an Asian Trinidadian Canadian man, speaking in the voice of a Japanese woman: the writer, and his grateful readers, simply refused to be contained by the limits the world regards as normal. This is the immense power that literature puts in the hands of all of us.

In the same way, literature offers us the opportunity to escape the two most pressing forms of bondage in our normal existence: time and ego. Emotionally and intellectually, literature dissolves the rules of time and beckons us toward Periclean Athens, Czarist Russia,

Elizabethan England, and a thousand other moments in the past. By lengthening our sense of time, it saves us from the maddening urgencies of the present. And when it succeeds on the highest level, it breaks the shell of our intense and tiresome self-consciousness. It forces itself inside the egotism fostered by the pressures of our lives and links us with human history and the vast ocean of humanity now on Earth. By taking us into other lives, it deepens our own.

Our clear task, if we hope to realize ourselves as a civilization, is to cherish the writers who have done their work and nourish the writers who are still doing it. The literary imagination is not a grace of life or a diversion: it is the best way we have found of reaching for the meaning of existence.

From the
FIFTEENTH
District

MAVIS GALLANT

ALTHOUGH AN EPIDEMIC OF HAUNTING, WIDELY reported, spread through the Fifteenth District of our city last summer, only three acceptable complaints were lodged with the police.

Major Emery Travella, 31st Infantry, 1914-1918, Order of the Leopard, Military Beech Leaf, Cross of St. Lambert First Class, killed while defusing a bomb in a civilian area 9 June 1941, Medal of Danzig (posthumous), claims he is haunted by the entire congregation of St. Michael and All Angels on Bartholomew Street. Every year on the Sunday falling nearest the anniversary of his death, Major Travella attends Holy Communion service at St. Michael's, the church from which he was buried. He stands at the back, close to the doors, waiting until all the communicants have returned to their places, before he approaches the altar rail. His intention is to avoid a mixed queue of dead and living, the thought of which is disgusting to him. The congregation sits, hushed and expectant, straining to hear the Major's footsteps (he drags one foot a little). After receiving the host, the Major leaves at once, without waiting for the Blessing. For the past several years, the Major has noticed that the congregation doubles in size as 9 June approaches. Some of these strangers bring cameras and tape recorders with them; others burn incense under the pews and wave amulets and trinkets in what they imagine to be his direction, muttering pagan gibberish all the while. References he is sure must be meant for him are worked in to the sermons: "And he that was dead sat up, and

Mavis Gallant was born in Montreal in 1922 and has lived in Paris since 1950. She is an officer of the Order of Canada and a winner of the Governor General's Award. Contrasting real and invented landscapes, she writes, "Against the sustained tick of the watch, fiction takes the measure of a life, a season, a look exchanged, the turning point, desire as brief as a dream, the grief and the terror that after childhood we cease to express. The lie, the look, the grief are without permanence. The watch continues to tick when the story stops."

began to speak" (Luke 7:15), or "So Job died, being old and full of days" (Job 42:17). The Major points out that he never speaks and never opens his mouth except to receive Holy Communion. He lived about sixteen thousand and sixty days, many of which he does not remember. On 23 September, 1914, as a young private, he was crucified to a cart wheel for five hours for having failed to salute an equally young lieutenant. One ankle was left permanently impaired.

The Major wishes the congregation to leave him in peace. The opacity of the living, their heaviness and dullness, the moisture of their skin, and the dustiness of their hair are repellent to a man of feeling. It was always his habit to avoid civilian crowds. He lived for six years on the fourth floor in Block E, Stoneflower Gardens, without saying a word to his neighbours or even attempting to learn their names. An affidavit can easily be obtained from the former porter at the Gardens, now residing at the Institute for Victims of Senior Trauma, Fifteenth District.

Mrs. Ibrahim, aged thirty-seven, mother of twelve children, complains about being haunted by Dr. L. Chalmeton of Regius Hospital, Seventh District, and by Miss Alicia Fohrenbach, social investigator from the Welfare Bureau, Fifteenth District. These two haunt Mrs. Ibrahim without respite, presenting for her ratification and approval conflicting and unpleasant versions of her own death.

According to Dr. Chalmeton's account, soon after Mrs. Ibrahim was discharged as incurable from Regius Hospital he paid his patient a professional call. He arrived at a quarter past four on the first Tuesday of April, expecting to find the social investigator, with whom he had a firm appointment. Mrs. Ibrahim was discovered alone, in a windowless room, the walls of which were coated with whitish fungus a quarter of an inch thick, which rose to a height of about forty inches from the floor. Dr. Chalmeton inquired, "Where is the social investigator?" Mrs. Ibrahim pointed to her throat, reminding him that she could not reply. Several dark-eyed children peeped into the room and ran away. "How many are yours?" the doctor asked. Mrs. Ibrahim indicated six twice with her fingers. "Where do they sleep?" said the Doctor. Mrs. Ibrahim indicated the floor. Dr. Chalmeton said, "What does your husband do for a living?" Mrs. Ibrahim pointed to a workbench on which the doctor saw several pieces of finely wrought jewellery; he thought it a waste that skilled work had been lavished on what seemed to be plastics and base metals. Dr. Chalmeton made the patient as comfortable as he could, explaining that he could not administer drugs for the relief of pain until the social investigator had signed a receipt for them. Miss Fohrenbach arrived at five o'clock. It had taken her forty minutes to find a suitable parking space: the street appeared to be poor, but everyone living on it owned one or two cars. Dr. Chalmeton, who was angry at having been kept waiting, declared he would not be responsible for the safety of his patient in a room filled with mould. Miss Fohrenbach retorted that the District could not resettle a family of fourteen persons who were foreign-born when there was a long list of native citizens waiting for accommodation. Mrs. Ibrahim had in any case relinquished her right to a domicile in the Fifteenth District the day she lost consciousness in the road and allowed an ambulance to transport her to a hospital in the Seventh. It was up to the hospital to look after her now. Dr. Chalmeton pointed out that housing of patients is not the business of hospitals. It was well known that the foreign poor preferred to crowd together in the Fifteenth, where they could sing and dance in the streets and attend one another's weddings. Miss Fohrenbach declared that Mrs. Ibrahim could easily have moved her bed into the kitchen, which was warmer and which boasted a window. When Mrs. Ibrahim died, the children would be placed in foster homes, eliminating the need for a larger apartment. Dr. Chalmeton remembers Miss Fohrenbach's then crying, "Oh, why do all these people come here, where nobody wants them?" While he was trying to think of an answer, Mrs. Ibrahim died.

In her testimony, Miss Fohrenbach recalls that she had to beg and plead with Dr. Chalmeton to visit Mrs. Ibrahim, who had been discharged from Regius Hospital without medicines or prescriptions or advice or instructions. Miss Fohrenbach had returned several times that April day to see if the Doctor had arrived. The first thing Dr. Chalmeton said on entering the room was, "There is no way of helping these people. Even the simplest rules of hygiene are too complicated for them to follow. Wherever they settle, they spread disease and vermin. They have been responsible for outbreaks of aphthous stomatitis, hereditary hypoxia, coccidioidomycosis, gonorrheal arthritis, and scleroderma. Their eating habits are filthy.

They never wash their hands. The virus that attacks them breeds in dirt. We took in the patient against all rules, after the ambulance driver left her lying in the courtyard and drove off without asking for a receipt. Regius Hospital was built and endowed for ailing Greek scholars. Now it is crammed with unteachable persons who cannot read or write." His cheeks and forehead were flushed, his speech incoherent and blurred. According to the social investigator, he was the epitome of the broken down, irresponsible old rascals the Seventh District employs in its public services. Wondering at the effect this ranting of his might have on the patient, Miss Fohrenbach glanced at Mrs. Ibrahim and noticed she had died.

Mrs. Ibrahim's version of her death has the social investigator arriving first, bringing Mrs. Ibrahim a present of a wine-coloured dressing gown made of soft, quilted silk. Miss Fohrenbach explained that the gown was part of a donation of garments to the needy. Large plastic bags, decorated with a moss rose, the emblem of the Fifteenth District, and bearing the words "Clean Clothes for the Foreign-Born," had been distributed by volunteer workers in the more prosperous streets of the District. A few citizens kept the bags as souvenirs, but most had turned them in to the Welfare Bureau filled with attractive clothing, washed, ironed, and mended, and with missing buttons replaced. Mrs. Ibrahim sat up and put on the dressing gown, and the social investigator helped her button it. Then Miss Fohrenbach changed the bed linen and pulled the bed away from the wall. She sat down and took Mrs. Ibrahim's hand in hers and spoke about a new, sunny flat containing five warm rooms which would soon be available. Miss Fohrenbach said that arrangements had been made to send the twelve Ibrahim children to the mountains for special winter classes. They would be taught history and languages and would learn to ski.

The Doctor arrived soon after. He stopped and spoke to Mr. Ibrahim, who was sitting at his workbench making an emerald patch box. The Doctor said to him, "If you give me your social-security papers, I can attend to the medical insurance. It will save you a great deal of trouble." Mr. Ibrahim answered, "What is social security?" The Doctor examined the patch box and asked Mr. Ibrahim what he earned. Mr. Ibrahim told him, and the Doctor said, "But that is less than the minimum wage." Mr. Ibrahim said, "What is a minimum wage?" The Doctor turned to Miss Fohrenbach, saying, "We really must try and help them." Mrs. Ibrahim died. Mr. Ibrahim, when he understood that nothing could be done, lay face down on the floor, weeping loudly. Then he remembered the rules of hospitality and got up and gave each of the guests a present — for Miss Fohrenbach a belt made of Syriac coins, a copy of which is in the Cairo Museum, and for the Doctor a bracelet of precious metal engraved with pomegranates, about sixteen pomegranates in all, that has lifesaving properties.

Mrs. Ibrahim asks that her account of the afternoon be registered with the police as the true version and that copies be sent to the Doctor and the social investigator, with a courteous request for peace and silence.

Mrs. Carlotte Essling, née Holmquist, complains of being haunted by her husband, Professor Augustus Essling, the philosopher and historian. When they were married, the former Miss Holmquist was seventeen. Professor Essling, a widower, had four small children. He explained to Miss Holmquist why he wanted to marry again. He said, "I must have one person, preferably female, on whom I can depend absolutely, who will never betray me even in her thoughts. A disloyal thought revealed, a betrayal even in fantasy, would be enough to destroy me. Knowing that I may rely upon some one person will leave me free to continue my work without anxiety or distraction." The work was the Professor's lifelong examination of the philosopher Nicolas de Malebranche, for whom he had named his eldest child. "If I cannot have the unfailing loyalty I have described, I would as soon not marry at all," the Professor added. He had just begun work on *Malebranche and Materialism.*

Mrs. Essling recalls that at seventeen this seemed entirely within her possibilities, and she replied something like "Yes, I see," or "I quite understand," or "You needn't mention it again."

Mrs. Essling brought up her husband's four children and had two more of her own, and died after thirty-six years of marriage at the age of fifty-three. Her husband haunts her with proof of her goodness. He tells people that Mrs. Essling was born an angel, lived like an angel, and is an angel in eternity. Mrs. Essling would like relief from this charge. "Angel" is a loose way of speaking. She is astonished that the Professor cannot be more precise. Angels are created, not born. Nowhere in any written testimony will you find a scrap of proof that angels are "good." Some are merely messengers, others have a paramilitary function. All are stupid.

After her death, Mrs. Essling remained in the Fifteenth District. She says she can go nowhere without being accosted by the Professor, who, having completed the last phase of his work *Malebranche and Mysticism*, roams the streets, looking in shopwindows, eating lunch twice, in two different restaurants, telling his life story to waiters and bus drivers. When he sees Mrs. Essling, he calls out, "There you are!" and "What have you been sent to tell me?" and "Is there a message?" In July, catching sight of her at the open-air fruit market on Dulac Street, the Professor jumped off a bus, upsetting barrows of plums and apricots, waving an umbrella as he ran. Mrs. Essling had to take refuge in the cold-storage room of the central market, where, years ago, after she had ordered twenty pounds of raspberries and currants for making jelly, she was invited by the wholesale fruit dealer, Mr. Lobrano, aged twenty-nine, to spend a holiday with him in a charming southern city whose Mediterranean Baroque churches he described with much delicacy of feeling. Mrs. Essling was too startled to reply. Mistaking her silence, Mr. Lobrano then mentioned a northern city containing a Gothic cathedral. Mrs. Essling said that such a holiday was impossible. Mr. Lobrano asked for one good reason. Mrs. Essling was at that moment four months pregnant with her second child. Three stepchildren waited for her out in the street. A fourth stepchild was at home looking after the baby. Professor Essling, working on his *Malebranche and Money,* was at home, too, expecting his lunch. Mrs. Essling realized she could not give Mr. Lobrano one good reason. She left the cold-storage room without another word and did not return to it in her lifetime.

Mrs. Essling would have liked to be relieved of the Professor's gratitude. Having lived an exemplary life is one thing; to have it thrown up at one is another. She would like the police to send for Professor Essling and tell him so. She suggests that the police find some method of keeping him off the streets. The police ought to threaten him; frighten him; put the fear of the Devil into him. Philosophy has made him afraid of dying. Remind him about how he avoided writing his *Malebranche and Mortality*. He is an old man. It should be easy.

A
WRITER'S
freedom

NADINE GORDIMER

(1975)[1]

WHAT IS A WRITER'S FREEDOM?

To me it is his right to maintain and publish to the world a deep, intense, private view of the situation in which he finds his society. If he is to work as well as he can, he must take, and be granted, freedom from the public conformity of political interpretation, morals, and tastes.

Living when we do, where we do, as we do, "Freedom" leaps to mind as a political concept exclusively — and when people think of freedom for writers they visualize at once the great mound of burnt, banned and proscribed books our civilization has piled up; a pyre to which our own country has added and is adding its contribution. The right to be left alone to write what one pleases is not an academic issue to those of us who live and work in South Africa. The private view always has been and always will be a source of fear and anger to proponents of a way of life, such as the white man's in South Africa, that does not bear looking at except in the light of a special self-justificatory doctrine.

All that the writer can do, as a writer, is to go on writing *the truth as he sees it.* That is what I mean by his "private view" of events, whether they be the great public ones of wars and revolutions, or the individual and intimate ones of daily, personal life.

As to the fate of his books — there comes a time in the history of certain countries when the feelings of their writers are best expressed in this poem, written within the lifetime of many of us, by Bertolt Brecht:

Nadine Gordimer was born in 1923 at Springs, Transvaal, near Johannesburg in South Africa and won the Nobel Prize for Literature in 1991. Her books have been banned over the years in her homeland because of her condemnation of apartheid. About such experiences, she reflects, "The only dictum I always remember is Andre Gide's — 'Salvation, for the writer, lies in being sincere even against one's better judgement.'"

When the Regime ordered that books with dangerous
    teachings
Should be publicly burnt and everywhere
Oxen were forced to draw carts full of books
To the funeral pyre, an exiled poet,
One of the best, discovered with fury, when he studied the
    list
Of the burned, that his books
Had been forgotten. He rushed to his writing table
On wings of anger and wrote a letter to those in power.
Burn me, he wrote with hurrying pen, burn me!
Do not treat me in this fashion. Don't leave me out. Have I
    not
Always spoken the truth in my books? And now
You treat me like a liar! I order you:
Burn me![2]

We South African writers can understand the desperate sentiments expressed while still putting up the fight to have our books read rather than burnt.

Bannings and banishments are terrible known hazards a writer must face, and many have faced, if the writer belongs where freedom of expression, among other freedoms, is withheld, but sometimes creativity is frozen rather than destroyed. A Thomas Mann survives exile to write a *Doctor Faustus*; a Pasternak smuggles *Doctor Zhivago* out of a ten-year silence; a Solzhenitsyn emerges with his terrible world intact in the map of *The Gulag Archipelago*; nearer our home continent, a Chinua Achebe, writing from America, does not trim his prose to please a Nigerian regime under which he cannot live;[3] a Dennis Brutus grows in reputation abroad while his poetry remains forbidden at home; and a Breyten Breytenbach, after accepting the special dispensation from a racialist law which allowed him to visit his home country with a wife who is not white, no doubt accepts the equally curious circumstance that the book he was to write about the visit was to be banned, in due course.[4]

Through all these vicissitudes, real writers go on writing the truth as they see it. And they do not agree to censor themselves...

You can burn the books, but the integrity of creative artists is not incarnate on paper any more than on canvas — it survives so long as the artist himself cannot be persuaded, cajoled or frightened into betraying it.

All this, hard though it is to live, is the part of the writer's fight for freedom the *world* finds easiest to understand.

There is another threat to that freedom, in any country where political freedom is withheld. It is a more insidious one, and one of which fewer people will be aware. It's a threat which comes from the very strength of the writer's opposition to repression of political freedom. That other, paradoxically wider, composite freedom — the freedom of his private view of life — may be threatened by the very awareness of *what is expected of him*. And often what is expected of him is conformity to an orthodoxy of opposition.

There will be those who regard him as their mouth-piece; people whose ideals, as a human being he shares, and whose cause, as a human being, is his own. They may be those whose suffering is his own. His identification with, admiration for, and loyalty to these set up a state of conflict within him. His integrity as a human being demands the sacrifice of everything to the struggle put up on the side of free men. His integrity as a writer goes the moment he begins to write what he is told he ought to write.

This is — whether all admit it or not — and will continue to be a particular problem for black writers in South Africa. For them, it extends even to an orthodoxy of vocabulary: the jargon of struggle, derived internationally, is right and adequate for the public platform, the newsletter, the statement from the dock; it is not adequate, it is not deep enough, wide enough, flexible enough, cutting enough, fresh enough for the vocabulary of the poet, the short story writer, or the novelist.

Neither is it, as the claim will be made, "a language of the people" in a situation where certainly it is very important that imaginative writing must not reach an elite only. The jargon of struggle lacks both the inventive pragmatism and the poetry of common speech — those qualities the writer faces the challenge to capture and explore imaginatively, expressing as they do the soul and identity of a people as no thousandth-hand "noble evocation" of clichés ever could.

The black writer needs his freedom to assert that the idiom of Chatsworth, Dimbaza, Soweto is no less a vehicle for the expression of pride, self-respect, suffering, anger — or anything else in the spectrum of thought and emotion — than the language of Watts or Harlem.

The fact is, even on the side of the angels, a writer has to reserve the right to tell the truth as he sees it, in his own words, without being accused of letting the side down. For as Philip Toynbee has written, "the writer's gift to the reader is not social zest or moral improvement or love of country, but an enlargement of the reader's apprehension."

This is the writer's unique contribution to social change. He needs to be left alone, by brothers as well as enemies, to make this gift. And he must make it even against his own inclination.

I need hardly add this does not mean he retreats to an ivory tower. The gift cannot be made from any such place. The other day, Jean-Paul Sartre gave the following definition of the writer's responsibility to his society as an intellectual, after himself having occupied such a position in France for the best part of seventy years: "He is someone who is faithful to a political and social body but never stops contesting it. Of course, a contradiction may arise between his fidelity and his *contestation,* but that's a fruitful contradiction. If there's fidelity without *contestation,* that's no good: one is no longer a free man."

When a writer claims these kinds of freedom for himself he begins to understand the real magnitude of his struggle. It is not a new problem and of all the writers who have had to face it, I don't think anyone has seen it as clearly or dealt with it with such uncompromising honesty as the great nineteenth-century Russian, Ivan Turgenev. Turgenev had an immense reputation as a progressive writer. He was closely connected with the progressive movement in Czarist Russia and particularly with its more revolutionary wing headed by the critic Belinsky and afterwards by the poet Nekrasov. With his sketches and stories, people said that Turgenev was continuing the work Gogol had begun of awakening the conscience of the educated classes in Russia to the evils of a political regime based on serfdom.

But his friends, admirers and fellow progressives stopped

short, in their understanding of his genius, of the very thing that made him one — his scrupulous reserve of the writer's freedom to reproduce truth and the reality of life even if this truth does not coincide with his own sympathies.

When his greatest novel, *Fathers and Sons*, was published in 1862, he was attacked not only by the right for pandering to the revolutionary nihilists, but far more bitterly by the left, the younger generation themselves, of whom his chief character in the novel, Bazarov, was both prototype and apotheosis. The radicals and liberals, among whom Turgenev himself belonged, lambasted him as a traitor because Bazarov was presented with all the faults and contradictions that Turgenev saw in his own type, in himself, so to speak, and whom he created as he did because — in his own words — "in the given case, life happened to be like that."

The attacks were renewed after the publication of another novel, *Smoke,* and Turgenev decided to write a series of autobiographical reminiscences which would allow him to reply to his critics by explaining his views on the art of writing, the place of the writer in society, and what the writer's attitude to the controversial problems of his day should be. The result was a series of unpretentious essays that make up a remarkable testament to a writer's creed. Dealing particularly with Bazarov and *Fathers and Sons*, he wrote of his critics:

> ... generally speaking (they) have not got quite the right idea of what is taking place in the mind of an author or what exactly his joys and sorrows, his aims, successes and failures are. They do not, for instance, even suspect the pleasure which Gogol mentions and which consists of castigating oneself and one's faults in the imaginary characters one depicts; they are quite sure that all a writer does is to "develop his ideas"... Let me illustrate my meaning by a small example. I am an inverterate and incorrigible Westerner. I have never concealed it and I am not concealing it now. And yet in spite of that it has given me great pleasure to show up in the person of Panshin (in *A House of Gentlefolk*) all the common and vulgar sides of the Westerners: I made the Slavophil Lavretsky "crush him utterly." Why did I do it,

I who consider the Slavophil doctrine false and futile? Because, *in the given case, life, according to my ideas, happened to be like that*, and what I wanted above all was to be sincere and truthful. In depicting Bazarov's personality, I excluded everything artistic from the range of his sympathies, I made him express himself in harsh and unceremonious tones, not out of an absurd desire to insult the younger generation ... but simply as a result of my observations ... My personal predilections had nothing to do with it. But I expect many of my readers will be surprised if I tell them that with the exception of Bazarov's views on art, I share almost all his convictions.[5]

And in another essay, Turgenev sums up regarding what he calls "the man of real talent": "The life that surrounds him provides him with the contents of his works; he is its concentrated reflection; but he is as incapable of writing a panegyric as a lampoon ... When all is said and done — that is beneath him. Only those who can do no better submit to a given theme or carry out a program."[6]

These conditions about which I have been talking are the special, though common ones of writers beleaguered in the time of the bomb and the colour-bar, as they were in the time of the jack-boot and rubber truncheon, and, no doubt, back through the ages whose shameful symbols keep tally of oppression in the skeleton cupboard of our civilizations.

Other conditions, more transient, less violent, affect the freedom of a writer's mind.

What about literary fashion, for example? What about the cycle of the innovator, the imitators, the debasers, and then the bringing forth of an innovator again? A writer must not be made too conscious of literary fashion, anymore than he must allow himself to be inhibited by the mandarins if he is to get on with work that is his own. I say "made conscious" because literary fashion is a part of his working conditions; he can make the choice of rejecting it, but he cannot choose whether it is urged upon him or not by publishers and readers, who do not let him forget he has to eat.

That rare marvel, an innovator, should be received with shock and excitement. And his impact may set off people in new directions of their own. But the next innovator rarely, I would almost say never, comes from his imitators, those who create a fashion in his image. Not all worthwhile writing is an innovation, but I believe it always comes from an individual vision, privately pursued. The pursuit may stem from a tradition, but a tradition implies a choice of influence, whereas a fashion makes the influence of the moment the only one for all who are contemporary with it.

A writer needs all these kinds of freedom, built on the basic one of freedom from censorship. He does not ask for shelter from living, but for exposure to it without possibility of evasion. He is fiercely engaged with life on his own terms, and ought to be left to it, if anything is to come of the struggle. Any government, any society — any vision of a future society — that has respect for its writers must set them as free as possible to write in their own various ways in their own choices of form and language, and according to their own discovery of truth.

Again, Turgenev expresses this best: "without freedom in the widest sense of the word — in relation to oneself... indeed, to one's people and one's history — a true artist is unthinkable; without that air it is impossible to breathe."[7]

And I add my last word: In that air alone, commitment and creative freedom become one.

---

1 Address, given at the Durban Indian Teachers' Conference, December 1975. First published in *New Classic*, no. 2 (1975), pp. 11-16.

2 Gordimer's source: "Die Büchverbrennung," translated by H.R. Hays as "The Burning of the Books," in *Bertolt Brecht, Selected Poems* (New York: Grove Press, 1959), p. 125.

3 Chinua Achebe, famous for his set of novels beginning with *Things Fall Apart* (London: Heinemann, 1958; New York: McDowell, Obolensky, 1959), was involved with the Biafran cause during the Nigerian Civil War, and was in exile in the United States at the time of this address. Gordimer reviewed his collection of essays, *Morning Yet on Creation Day* (London: Heinemann, 1975) in the *Times Literary Supplement*, 17 October 1975, p. 1227.

4 Breyten Breytenbach, most brilliant Afrikaans writer of the Sestiger ("Sixties") generation, which for the first time began to challenge Afrikaner culture in a dramatic way. While living in Paris he married a Vietnamese woman; this was

illegal under the South African Prohibition of Mixed Marriages Act, which was why he needed special dispensation to return with his wife. His book about the visit was *'n Seisoen in die Paradys*, first published in Afrikaans under the pseudonym of B.B. Lasarus (Johannesburg: Perskor, 1976), and later under Breytenbach's own name as *A Season in Paradise*, translated by Rike Vaughan, introduction by André Brink (New York: Persea, 1980; London: Faber, 1985). See "The Essential Gesture," n. 7 below.

5 From Ivan Turgenev, "Apropos of Father and Sons," in Turgenev's *Literary Reminiscences and Autobiographical Fragments,* translated with an introduction by David Magarshack, and an essay by Edmund Wilson (London: Faber, 1958), pp. 170-1.

6 From Turgenev's introduction to the collected edition of his novels, quoted by Magarshack in his introduction, ibid, p. 82.

7 "Apropos of *Fathers and Sons*," p, 176.

Summer
MEDITATIONS

VÁCLAV HAVEL

TODAY WE OFTEN HEAR THE LINE, "WE needn't discover what has already been discovered! Why reinvent the wheel?" I understand this sentiment and I fully agree with it — most of the time. Indeed, it makes no sense to attempt to rediscover the law of supply and demand, the principle of shareholding or value-added tax, the basic constellation of human rights and freedoms, techniques of municipal self-government, tried and true elements of parliamentary democracy, or lead-free gasoline.

I even understand this sentiment when it means something broader, a more general message that might be formulated as: "Let us be done with the silly, inflated notion that Czechoslovakia is the navel of the world, capable of endowing humanity with a brand-new and unheard-of political and economic system, one that will take the world by storm." If the sentiment is a protest against the conceited idea that we alone are capable of inventing a better world,

Václav Havel was born in Czechoslovakia in 1936. He has written many plays and influential essays on the nature of totalitarianism and dissent. A book of letters to his wife, Olga, provides an account of his imprisonment for his involvement in the Czech human rights movement. In December 1989, he was elected president of Czechoslovakia. He writes: "The real reason I am always creating something, organizing something ... is to vindicate my permanently questionable right to exist.... It's a paradox, but I must admit that if I am a better president than many others would be in my place, then it is precisely because somewhere in the deepest substratum of my work lies this constant doubt about myself and my right to hold office."

then again I can only concur. After all, it was I who long ago, back in 1968, made a lot of enemies by ridiculing the illusions of reform Communists that we were practically the most important country in the world because we were the first to try to combine socialism and democracy. These days, such objections are aimed at advocates of the so-called third way, which is meant to be some combination of capitalism and socialism. I don't know exactly how anyone understands this "third way," in specific terms, but if it is meant to refer to some combination of the proven and the unproven, I must place myself on the side of those who would rather not have anything to do with it.

But sometimes — especially in the hands of people with a tendency towards dogmatic, ideological thinking — this sentiment becomes a kind of hickory stick to crack across the knuckles of anyone who does not want, for whatever reason, to copy faithfully all the models presented — which today, of course, are Western models. If that is what it means, then I can't agree. Without being, as I have said, a seeker after some "third way," I am opposed to blind imitation, especially if it becomes an ideology. My reason for this is very simple: it is against nature and against life. We will never turn Czechoslovakia into a Federal Republic of Germany, or a France, or a Sweden, or a United States of America, and I don't see the slightest reason why we should try. That would only raise the question of why we should be an independent country at all. Why bother learning such unimportant languages as Czech and Slovak in school? Why not apply at once to be the fifty-first state of the USA?

Life and the world are as beautiful and interesting as they are because, among other things, they are varied, because every living creature, every community, every country, every nation has its own unique identity. France is different from Spain and Spain is not the same as Finland. Each country has its own geographical, social, intellectual, cultural, and political climate. It is proper that things should be this way, and I cannot understand why we alone should be so ashamed of ourselves that we don't want to be Czechoslovakia. To me, this is like going from one extreme to another: one moment we take on the role of a world messiah; the next we are deeply ashamed of our very existence. (This, of course, is nothing new: we have experienced these swings from pomposity to masochism and back many times.)

To sum up: though we haven't the slightest reason not to learn from any place in the world that can offer us useful knowledge, at the same time I see no reason why we should be ashamed of trying to find our own way, one that derives from our Czechoslovakia identity. In many cases we haven't really any choice. We are not doing this to dazzle the world with our originality, or to cure some inferiority complex. We are doing it purely and simply because it is the only possible way: our country is where it is, its landscape is beautiful in certain ways and devastated in others, its natural resources and industries are structured in such-and-such a way, we speak the languages we speak, we have our own historical traditions and customs, the political right and left are the way they are here and not the way they are elsewhere, and no matter how much we might want to, we can scarcely hope to change these things entirely. Why not try to accept all this as fact? Why not try to understand the inner content of this fact, the potential, the problems and hopes connected with it? And why not deal with it in the most appropriate and adequate way?

Having said that we must build a state based on intellectual and spiritual values, I must now touch on the question of what our intellectual and spiritual potential is, and whether it has any distinctive features at all. But to do that, I had first to come to terms with the possible accusation that I was seeking for our country something as shameful as its "own way."

Yes, our intellectual and spiritual potential really does have its own identity. We are what history has made us. We live in the very centre of Central Europe, in a place that from the beginning of time has been the main European crossroads of every possible interest, invasion, and influence of a political, military, ethnic, religious, or cultural nature. The intellectual and spiritual currents of east and west, north and south, Catholic and Protestant, enlightened and romantic — the political movements of conservative and progressive, liberal and socialist, imperialist and national liberationist — all of these overlapped here, and bubbled away in one vast cauldron, combining to form our national and cultural consciousness, our traditions, the social models of our behaviour, which have been passed down from generation to generation. In short, our history has formed our experience in the world.

For centuries, we — Czechs and Slovaks, whether in our own state or under foreign control — lived in a situation of constant menace from without. We are like a sponge that has gradually absorbed and digested all kinds of intellectual and cultural impulses and initiatives. Many European initiatives were born or first formulated here. At the same time, our historical experience has imbued us with a keen sensitivity to danger, including danger on a global scale. It has even made us somewhat prescient: many admonitory visions of the future — Kafka's and Capek's for instance — have come from here — and not by chance. The ethnic variety of this area, and life under foreign hegemony, have created different mutations of our specific Central European provinciality, which have frequently, and in very curious ways, merged with that clairvoyance.

Our most recent great experience, an experience none of the Western democracies has ever undergone, was Communism. Often we ourselves are unable to appreciate fully the existential dimension of this bitter experience and all its consequences, including those that are entirely metaphysical. It is up to us alone to determine what value we place on that particular capital.

It is no accident that here, in this milieu of unrelenting danger, with the constant need to defend our own identity, the idea that a price must be paid for truth, the idea of truth as a moral value, has such a long tradition. That tradition stretches from Saints Cyril and Methodius, who brought Christianity to the region in the ninth century A.D., through the fifteenth-century reformer Jan Hus, all the way down to modern politicians like Tomás Garrigue Masaryk and Milan Stefánik, and the philosopher Jan Patocka.

When we think about all this, the shape of our present intellectual and spiritual character starts to appear — the outlines of an existential, social, and cultural potential which is slumbering here and which — if understood and evaluated — can give the spirit, or the idea, of our new state a unique and individual face.

Every European country has something particular to it — and that makes its autonomy worth defending, even in the framework of an integrating Europe. That autonomy then enriches the entire European scene; it is another voice in that remarkable polyphony, another instrument in that orchestra. And I feel that our historical experience, our intellectual and spiritual potential, our experience of misery, absurdity, violence, and idyllic tranquillity, our humour, our experience of sacrifice, our love of civility, our love of truth and our knowledge of the many ways truth can be betrayed — all this can, if we wish, create another of those distinct voices from which the chorus of Europe is composed.

We must learn wherever we can. But we can also offer something: not only the inimitable climate of our mind and spirit, not only the message we have mined from our historical experience, but — God willing — perhaps even an original way of breathing this character and experience into the newly laid foundations of our state, into the architecture of its institutions and the features of its culture.

Our great, specific experience of recent times is the collapse of an ideology. We have all lived through its tortured and complicated vagaries, and we have gone through it, as it were, to the bitter end. This experience has, to an extraordinary degree, strengthened my ancient scepticism towards all ideologies. I think that the world of ideologies and doctrines is on the way out for good — along with the entire modern age. We are on the threshold of an era of globality, an era of open society, an era in which ideologies will be replaced by ideas.

Building an intellectual and spiritual state — a state based on ideas — does not mean building an ideological state. Indeed, an ideological state cannot be intellectual or spiritual. A state based on ideas is precisely the opposite: it is meant to extricate human beings from the straitjacket of ideological interpretations, and to rehabilitate them as subjects of individual conscience, of individual thinking backed up by experience, of individual responsibility, and with a love for their neighbours that is anything but abstract.

A state based on ideas should be no more and no less than a guarantee of freedom and security for people who know that the state and its institutions can stand behind them only if they themselves take responsibility for the state — that is, if they see it as their own project and their own home, as something they need not fear, as something they can — without shame — love, because they have built it for themselves.

Translated by Paul Wilson

The
PRISONER
who wore glasses

BESSIE HEAD

Bessie Head was born in Pietermaritzburg, South Africa in 1937. The child of an illegal union of mixed parents, she was taken from her mother at birth, raised by foster parents until she was thirteen, and then placed in a mission orphanage. She eventually abandoned her homeland, her teaching career, and an early marriage, and sought asylum in a simple village life in Botswana. She described her stories as reflections of daily encounters with undistinguished people and yet found in stories "a world of magic beyond your own." She died in 1986.

SCARCELY A BREATH OF WIND DISTURBED THE stillness of the day and the long rows of cabbages were bright green in the sunlight. Large white clouds drifted slowly across the deep blue sky. Now and then they obscured the sun and caused a chill on the backs of the prisoners who had to work all day long in the cabbage field. This trick the clouds were playing on the sun eventually caused one of the prisoners who wore glasses to stop work, straighten up and peer short-sightedly at them. He was a thin little fellow with a hollowed-out chest and comic knobbly knees. He also had a lot of fanciful ideas because he smiled at the clouds.

"Perhaps they want me to send a message to the children," he thought, tenderly, noting that the clouds were drifting in the direction of his home some hundred miles away. But before he could frame the message, the warder in charge of his work span shouted: "Hey, what you tink you're doing, Brille?"

The prisoner swung round, blinking rapidly, yet at the same time sizing up the enemy. He was a new warder, named Jacobus Stephanus Hannetjie. His eyes were the colour of the sky but they were frightening. A simple, primitive, brutal soul gazed out of them. The prisoner bent down quickly and a message was quietly passed down the line: "We're in for trouble this time, comrades."

"Why?" rippled back up the line.

"Because he's not human," the reply rippled down and yet only the crunching of the spades as they turned over the earth disturbed the stillness.

This particular work span was known as Span One. It was composed of ten men and they were all political prisoners. They were grouped together for convenience as it was one of the prison regulations that no black warder should be in charge of a political prisoner lest this prisoner convert him to the views. It never seemed to occur to the authorities that this very reasoning was the strength of Span One and a clue to the strange terror they aroused in the warders. As political prisoners they were unlike the other prisoners in the sense that they felt no guilt nor were they outcasts of society. All guilty men instinctively cower, which was why it was the kind of prison where men got knocked out cold with a blow at the back of the head from an iron bar. Up until the arrival of Warder Hannetjie, no warder had dared beat any member of Span One and no warder had lasted more than a week with them. The battle was entirely psychological. Span One was assertive and it was beyond the scope of white warders to handle assertive black men. Thus, Span One had got out of control. They were the best thieves and liars in the camp. They lived all day on raw cabbages. They chatted and smoked tobacco. And since they moved, thought and acted as one, they had perfected every technique of group concealment.

Trouble began that very day between Span One and Warder Hannetjie. It was because of the shortsightedness of Brille. That was the nickname he was given in prison and is the Afrikaans word for someone who wears glasses. Brille could never judge the approach of the prison gates and on several previous occasions he had munched on cabbages and dropped them almost at the feet of the warder and all previous warders had overlooked this. Not so Warder Hannetjie.

"Who dropped that cabbage?" he thundered.

Brille stepped out of line.

"I did," he said meekly.

"All right," said Hannetjie. "The whole Span goes three meals off."

"But I told you I did it," Brille protested.

The blood rushed to Warder Hannetjie's face.

"Look 'ere," he said. "I don't take orders from a kaffir. I don't know what kind of kaffir you tink you are. Why don't you say Baas. I'm your Baas. Why don't you say Baas, hey?"

Brille blinked his eyes rapidly but by contrast his voice was strangely calm.

"I'm twenty years older than you," he said. It was the first thing that came to mind but the comrades seemed to think it a huge joke. A titter swept up the line. The next thing Warder Hannetjie whipped out a knobkerrie and gave Brille several blows about the head. What surprised his comrades was the speed with which Brille had removed his glasses or else they would have been smashed to pieces on the ground.

That evening in the cell Brille was very apologetic.

"I'm sorry, comrades," he said. "I've put you into a hell of a mess."

"Never mind, brother," they said. "What happens to one of us, happens to all."

"I'll try to make up for it, comrades," he said. "I'll steal something so that you don't go hungry."

Privately, Brille was very philosophical about his head wounds. It was the first time an act of violence had been perpetrated against him but he had long been a witness of extreme, almost unbelievable human brutality. He had twelve children and his mind travelled back that evening though the sixteen years of bedlam in which he had lived. It had all happened in a small drab little three-bedroomed house in a small drab little street in the Eastern Cape and the children kept coming year after year because neither he nor Martha ever managed the contraceptives the right way and a teacher's salary never allowed moving to a bigger house and he was always taking exams to improve his salary only to have it all eaten up by hungry mouths. Everything was pretty horrible, especially the way the children fought. They'd get hold of each other's heads and give them a good bashing against the wall. Martha gave up something along the line so they worked out a thing between them. The bashings, biting, and blood were to operate in full swing until he came home. He was to be the bogey-man and when it worked he never failed to have a sense of godhead at the way in which his presence could change savages into fairly reasonable human beings.

Yet somehow it was this chaos and mismanagement at the centre of his life that drove him into politics. It was really an ordered beautiful world with just a few basic slogans to learn along with the rights of mankind. At one stage, before things became very bad, there were conferences to attend, all very far away from home.

"Let's face it," he thought ruefully. "I'm only learning right now what it means to be a politician. All this while I've been running away from Martha and the kids."

And the pain in his head brought a hard lump to his throat. That was what the children did to each other daily and Martha wasn't managing and if Warder Hannetjie had not interrupted him that morning he would have sent the following message: "Be good comrades, my children. Cooperate, then life will run smoothly."

The next day Warder Hannetjie caught this old man of twelve children stealing grapes from the farm shed. They were an enormous quantity of grapes in a ten gallon tin and for this misdeed the old man spent a week in the isolation cell. In fact, Span One as a whole was in constant trouble. Warder Hannetjie seemed to have eyes at the back of his head. He uncovered the trick about the cabbages, how they were split in two with the spade and immediately covered with earth and then unearthed again and eaten with split-second timing. He found out how tobacco smoke was beaten into the ground and he found out how conversations were whispered down the wind.

For about two weeks Span One lived in acute misery. The cabbages, tobacco, and conversations had been the pivot of jail life to them. Then one evening they noticed that their good old comrade who wore the glasses was looking rather pleased with himself. He pulled out a four ounce packet of tobacco by way of explanation and the comrades fell upon it with great greed. Brille merely smiled. After all, he was the father of many children. But when the last shred had disappeared, it occurred to the comrades that they ought to be puzzled. Someone said: "I say, brother. We're watched like hawks these days. Where did you get the tobacco?"

"Hannetjie gave it to me," said Brille.

There was a long silence. Into it dropped a quiet bombshell.

"I saw Hannetjie in the shed today," and the failing eyesight blinked rapidly. "I caught him in the act of stealing five bags of fertilizer and he bribed me to keep my mouth shut."

There was another long silence.

"Prison is an evil life," Brille continued, apparently discussing some irrelevant matter. "It makes a man contemplate all kinds of evil deeds."

He held out his hand and closed it.

"You know, comrades," he said. "I've got Hannetjie. I'll betray him tomorrow."

Everyone began talking at once.

"Forget it, brother. You'll get shot."

Brille laughed.

"I won't," he said. "That is what I mean about evil. I am a father of children and I saw today that Hannetjie is just a child and stupidly truthful. I'm going to punish him severely because we need a good warder."

The following day, with Brille as witness, Hannetjie confessed to the theft of the fertilizer and was fined a large sum of money. From then on Span One did very much as they pleased while Warden Hannetjie stood by and said nothing. But it was Brille who carried this to extremes. One day, at the close of work Warder Hannetjie said: "Brille, pick up my jacket and carry it back to the camp."

"But nothing in the regulations says I'm your servant, Hannetjie," Brille replied coolly.

"I've told you not to call me Hannetjie. You must say Baas," but Warder Hannetjie's voice lacked conviction. In turn, Brille squinted up at him.

"I'll tell you something about this Baas business, Hannetjie," he said. "One of these days we are going to run the country. You are going to clean my car. Now, I have a fifteen year old son and I'd die of shame if you had to tell him that I ever called you Baas."

Warder Hannetjie went red in the face and picked up his coat.

On another occasion Brille was seen to be walking about the prison yard, openly smoking tobacco. On being taken before the prison commander he claimed to have received the tobacco from Warder Hannetjie. All throughout the tirade from his chief, Warder Hannetjie failed to defend himself but his nerve broke completely. He called Brille to one side.

"Brille," he said. "This thing between you and me must end. You may not know it but I have a wife and children and you're driving me to suicide."

"Why don't you like your own medicine, Hannetjie?" Brille asked quietly.

"I can give you anything you want," Warder Hannetjie said in desperation.

"It's not only me but the whole of Span One," said Brille, cunningly. "The whole of Span One wants something from you."

Warder Hannetjie brightened with relief.

"I think I can manage if it's tobacco you want," he said.

Brille looked at him, for the first time struck with pity, and guilt.

He wondered if he had carried the whole business too far. The man was really a child.

"It's not tobacco we want, but you," he said. "We want you on our side. We want a good warder because without a good warder we won't be able to manage the long stretch ahead."

Warder Hannetjie interpreted this request in his own fashion and his interpretation of what was good and human often left the prisoners of Span One speechless with surprise. He had a way of slipping off his revolver and picking up a spade and digging alongside Span One. He had a way of producing unheard of luxuries like boiled eggs from his farm nearby and things like cigarettes, and Span One responded nobly and got the reputation of being the best work span in the camp. And it wasn't only take from their side. They were awfully good at stealing certain commodities like fertilizer which were needed on the farm of Warder Hannetjie.

The Second
COMING
of Come-by-Chance

JANETTE TURNER HOSPITAL

Janette Turner Hospital was born in 1942 in Melbourne, Australia, travelled in England, America, and India, and has lived in Canada since 1971. She describes her characters as nomads "who cross borders, who straddle cultures and countries, who live with a constant sense of dislocation."

IN THE SIXTY-FOURTH MONTH OF the tribulation, just five weeks before the drought finally broke, people began driving out from Townsville and Ayr and Home Hill, from Charters Towers and Collinsville, and from any number of smaller salt-of-North-Queensland towns: Thalanga, Mungunburra, Millaroo, Mingela. The Flinders Highway was thick with four-wheel drives, the air with dust. Afterwards, newspapers remembered that there had been a curious sense of festivity about, a sort of overwrought camaraderie, the kind that comes in the wake of cyclones, earthquakes, bush fires. Post-traumatic hysteria, the articles said. Old men had visions, revenants appeared in the pubs, crackpots wrote to newspapers, children concocted secret ways of sucking juice from rocks and of finding the underground channels where the rivers had fled.

All this was mere prelude. It was Tom Kelly and Davy Cobb, unlikely angels of the apocalypse, who ushered in what the Brisbane papers dubbed a "Flight into Egypt" and the *Sydney Morning Herald*, predictably supercilious, headlined as "Latter Day Madness in Queensland." (Perhaps it is unnecessary, from this retrospective distance, and after so much analysis of the psychological effects of the drought, to note that those who live in the cities of the coastal plain, while not unaffected by years of water restrictions, are unlikely to be aware of the intensity of the inland thirst for something, for *anything,* to happen.) In any case, in the beginning it was just a trickle. Perhaps, that first weekend, only sixty people drove out to the dwindling Burdekin Dam to watch the reappearance of Come-by-Chance, for the tip of the Anglican steeple had been sighted by the two boys fishing in their homemade boat.

Sighted? *Bumped into* would be more accurate. Young Tom Kelly had laced a worm around his hook and cast his line. At the oars, Davy Cobb felt a jolt. What Tom hooked was the copper cross, green as verdigris, sticking out of the water like the index finger of God, potent, invisible (at least until the moment of reckoning). The faster Tom reeled in his catch, the swifter the little boat skimmed towards its ramming. Both boys went into the water like steeplejacks on the toss. This was a week before Christmas, and the momentum of Tom Kelly's unpremeditated dive was later likened by the Bishop of North Queensland to the downward swoop of the Incarnation. Tom claimed he looked through the rose window and saw a phosphorescent glow, then kept plummeting to the soft Gothic arch. The nave was full of green radiance.

"It was like there was sump'n *pullin'* me," he said. "I couldn't turn, I thought me lungs were gonna bust. Then I saw this kinda light, this kinda I dunno, like a million green parakeets' wings or sump'n, and then this blaze like a double-bunger star, it bloody well bursts inside me head. And next thing I know, Davy's thumpin' water outta me on the bank."

"An epiphany," the Bishop of North Queensland said. (It was the last Sunday in Advent.) But the pastor of the Gospel Hall in Mingela, the closest town to Come-by-Chance, thundered darkly: "And in those days there shall be signs and portents, for He shall come as a fire descending…"

It is reasonably safe to assume that the *Sydney Morning Herald* would not have mentioned this spiritual event had it not been for the impending election and the clear correlation between water levels in the Burdekin Dam and political chaos in Queensland. The drought, it will be recalled, at first confined to that arid crescent between Townsville and Mt. Isa, had spread like a virus. By the time of Tom Kelly's appearance on the front page of the tabloids, there were bush fires all the way to the Dandenongs and the Adelaide Hills. This "Queensland drift" seemed ominous, even to secular minds.

Addressing himself primarily to the political issues, a Sydney pundit commented, in passing, on the fishing story. Where else but in Queensland? he asked. Pressed by his interviewer to respond to a tabloid headline ("Christmas vision saves boy's life"), he spoke of the effects of shock and water-pressure and diminution of oxygen and concomitant hallucinatory indications such as the kind of aura that accompanies migraine or near-drowning, but no one in central or north Queensland watched this show. Indeed, even in Sydney and Melbourne, those infallible Geiger-counters of truth, many chose to ignore common sense. For what raconteur in the pub, what politician, what preacher, could resist Tom Kelly's aurora and the resurrection of Come-by-Chance?

"And there shall be famines," bishops and gospel firebrands read as with one voice. "And pestilences, and earthquakes in divers places. For then shall be great tribulation ...." On exegesis, however, the divines parted ways, and quite contradictory moral and political interpretations — not to mention voting admonitions — were brought to bear. It was only in the actual scriptural words of warning that they spoke as one again: "And except those days should be shortened, there should no flesh be saved: but for the elect's sake those days shall be shortened."

A swelling group of the elect began to gather for vigil and prayer and competitive political pamphleteering at the ever-lower waterline of the dam. A week after the fishing incident, the whole cross of St. Stephen Martyr was visible and a foot of steeple tiles below; in the second week, the belltower of the Catholic church appeared; in the third, the Post Office clock. Word spread along the stock routes and talk-show arteries, and via the pages that come round fish and chips. The elect were joined by the curious, the bored, the Sydney and Melbourne reporters, the television cameras, the signs-and-portents groupies. A camp was set up.

"What come ye out for to see?" the Mingela pastor, distributing tracts, asked through a megaphone. *Ahh, knock it off,* people said, but not too savagely. Long droughts of continental proportions induce nervous piety in many breasts — though not in all. Around the country, bookies also set up shop and punters laid bets on the next building to resurrect itself. Daily the odds were published on the likelihood of there still being skeletons anchored to the stools in the bar, because many people now recalled tales of Come-by-Chancers who had refused to leave town.

The "flight into Egypt" became a veritable exodus, and the *Sydney Morning Herald* ran a full weekend feature in which the word "mirage" was frequently mentioned. In Melbourne, the *Age* went as far as a reference to "collective hysteria." This was due to the curious fact that while everyone at the site, including visiting reporters, could clearly see the re-emergent town, no trace of it showed up in photographs. The Logos Foundation issued a statement to the press: *Faith is the substance of things hoped for, the evidence of things not seen.* Only the pure in heart, it was implied, can witness the unblemished city of God. Come-by-Chance became symbol and rallying cry for a lost way of life, a simpler cleaner time, which each political party vowed to restore.

In the capital cities, editors were deluged with letters. No one could have predicted the number of people still living who had visited, or had relatives in, or had themselves inhabited the town of Come-by-Chance before it went under the dam. By one newspaper's count, the population had been a quarter of a million just prior to inundation, though the town had boasted only three churches, a post office, seven pubs, a one-teacher school, a police station (with two constables assigned) and a handful of shops and houses. There was considerable divergence of opinion on the erstwhile economic base. Sheep, most claimed. Opal prospecting, others contended. Tall stories, suggested the literary editor of the *Australian*, a man noted for his scepticism and wit. He alluded to Ern Malley and the whole issue of the literary hoax. He quoted Banjo Paterson, and left readers to draw their own conclusions:

> *But my languid mood forsook me, when I found a*
> *name that took me;*
> *Quite by chance I came across it — "Come-by-*
> *Chance" was what I read;*
> *No location was assigned it, not a thing to help one*
> *find it,*
> *Just an N which stood for northward, and the rest*
> *was all unsaid ...*
> *But I fear, and more's the pity, that there's really*
> *no such city,*
> *For there's not a man can find it of the shrewdest*
> *folk I know;*

Janette Turner

> *"Come-by-Chance," be sure it never means a land*
> *of fierce endeavour —*
> *It is just the careless country where the dreamers*
> *only go.*

Or where the victims of nightmares are trapped, thought Mrs. Adeline Capper. *And they can never leave.*

Adeline Capper dreaded the newspapers and read them with a compulsive doomed fascination. She had always known there was no way of expunging the past. One could flee it, drown it, bury it, tear up the newsprint record, but it went on skulking around today. It was always *there.* Inside one. *Here.*

She was twenty then, sixty now, but twenty was as close as her skin.

It is the doing nothing that is intolerable, she thought. The fact that there is nothing to be done.

From her brown garden on the south side of Townsville, she watched the trekkers herding down the highway and out to the dam. A hot wind blew. Her bougainvillea made a dead parchment sound against the fence. Like a sleepwalker, she got into her car and followed the columns of dust.

A settlement — a tent city — had sprung up: trailers, kombi vans, canvas of all shapes and sizes, camp fires, styrofoam iceboxes full of beer. People told jokes and sang songs. There was raucous laughter, catcalls as Adeline Capper moved among them. Was she twenty or sixty? Night was such a dangerous time. *Hey, grandma!* someone called. *Wanna cuppa?*

"Hey, didn't mean to give you such a scare," a young man said, apologetic. "Here. Have a cuppa tea."

She took it, shivering in the dry evening heat. Her teeth chattered.

"You all right?" the young man's wife asked, concerned.

They always asked that, but no one wanted to know the true answer. No one ever wanted to know that. She couldn't remember how far the reserve was from here, how far she had walked.

"See?" the young wife asked. She held out a wooden plaque,

bleached colourless, soft to the touch. But you could still make out the carved indentation: XXXX.

"Four-X, the Queensland beer," the young man laughed. "A true-blue bit'v history. She was a bugger to rip off the wall, but."

"Brian dived," his wife said proudly. "You wouldn't believe what people are bringing up."

Oh, Adeline could well believe. And how soon would someone surface with her first teaching year? Who would wave it aloft? It was written in stone down there somewhere. Everywhere.

It was, it *is*, loud in the air.

"Where've you been, Adeline?" Sergeant Hobson had crooned, *croons*, pulling up beside her in the car. (Big Bob is what everyone calls him. She teaches his daughter in the one-teacher school.) He leans out and sings in her ear: "Oh where have you *been*, Adel-*een?*"

And she laughs, and Constable Terry Wilkes in the passenger seat laughs too. And they stand there in the moonlight on the dirt road winding into town from the reserve. "I've been visiting some of my school kids," she says. "The ones that live out ... uh, the Chillagong ones."

"The boongs, you mean?"

Boongs. Abos. Everyone says it. It strikes her as terribly rude, but she doesn't want to offend, doesn't want to appear stuck-up, doesn't want to sound like the smart-alec who just arrived from Brisbane. "Yeah," she says.

"Bit late, innit?" Big Bob asks. "To be walking back into town."

"Yeah, I reckon." She laughs nervously. (She'd been terrified as a matter of fact; and so relieved when the police car pulled up.) She'd thrown them into total confusion back there, Hazel, Evangeline, Joshua, their mothers and fathers, walking the three miles out after school. It isn't what anyone does, goes to Chillagong, she can see that now. "I got invited to stay for dinner." (Though she thinks they hadn't known what else to do, and nor had she; and the minute she'd accepted, she'd realized they hadn't expected her to. Perhaps hadn't wanted her to.)

"For *dinner*, well, stone the crows!" Big Bob and Constable Terry roar with laughter. "Kangaroo rat and witchetty grubs, ya like them, do ya?"

"Oh no," she shudders. In fact, she doesn't know what it was she ate. Some kind of stew.

"Well, get in then." Big Bob lumbers out and puts an arm around her shoulders. "We'll drive you back home." He presses his big fat lips against her neck and his beery breath hits her like a fist. "Can't have the little lady-teacher from Brisbane on her own in the bush at night." He strokes her hair protectively, and the front of her dress, accidentally pressing her breasts. She's a bit embarrassed, but they are the police, after all. She's between the two men on the Holden's bench-seat and feels safe.

"Adeline's been having a little night life with the boongs," Big Bob tells Constable Terry. "She likes those big fat witchetty grubs."

"Big fat witchetty grubs," Terry sings. "Oh Adeleen, our village queen, she loves those big fat witchetty grubs."

"No, no," she protests laughing, but Big Bob joins in, and they sing and sway and laugh in the dark, and after a while she sings along: "I love those big fat witchetty grubs."

"She loves those big black witchetty grubs," sings Constable Terry.

Both men laugh so much that the car slews onto the shoulder and back, then off the road again. She's nervous now. She thinks they are both quite drunk.

"You think they're better?" Big Bob demands. "The big black juicy ones?"

She doesn't know what to answer. "Where are we going?" she asks, alarmed.

There's a blur, both car doors opening, a blank.

What she remembers: spiky grass and ants against her skin, and words marching in ranks through her head. *I don't believe this, I don't believe it, it doesn't make sense, it isn't happening.* And then the next day (she must have slept, or been unconscious, whatever), the next day: Blood, bruises, and no clothes. No sign of her clothes. And diarrhea, the worst, the most humiliating thing.

But she can remember only grass and ants and the shapes of words. The words themselves are jagged, they hurt her skin. Fog comes and goes. A search party shouts, she hunches herself up, ashamed, ashamed. She doesn't want to be found. The pain from moving is so great that she blacks out.

"Christ!" Big Bob has tears in his eyes. He covers her with a blanket. The picture in the newspaper shows him cradling her in his arms. "We found her clothes on the reserve," Big Bob tells the reporter. "The animals won't get away with it, I can promise you that."

Days come and go. She's teaching again, it seems. The children stare and whisper, the reserve kids don't come any more. She cannot look at Margaret, Big Bob's daughter. When she walks into the general store, people fall silent. "Poor Addie," they murmur, as though she had a terminal disease. "At least the bastards are in gaol," someone says. She stares, puzzled; there is something just out of reach, but only words rattle in her head like small change in an empty tin can. She has a nightmare, and in the morning she forces herself to read the papers that have stacked themselves up, unopened. She sees Joshua's father, Evangeline's father, in handcuffs. *We didn't do it, boss. We dunno how her clothes ...*

A fever descends.

"Benevolent reasons" is what the transfer slip from the Education Department says. She lies awake all the last night, afraid. Wouldn't it be better, she begs herself, more sensible, to say nothing? Reasons for saying nothing marshall themselves in ranks, they file through her head all night. *Please,* her body begs. I have no choice, she tells it. I have to.

Morning. Two blocks to the police station, bodily panic, retreat. It takes her until the third attempt, and then Sergeant Big Bob Hobson and Constable Terry Wilkes greet her effusively. They take her into their office, they give her tea and a biscuit. "We're glad to see you up and about again," they say. "Glad to see you looking so well."

Her hands are sweating, her knees are weak, her throat dry.

"I am going to tell," she says. The noise each word makes as it falls on the floor is deafening.

They look at her blandly, innocent-eyed. "Tell what?" they ask.

Tell what? She feels dizzy, there is no bottom to this fall. She thinks: I will never know for sure again if night is night or day is day, what is dream or not-dream.

It would help, they told her at the hospital, to be a thousand miles away for a while. It would help, they said, to be somewhere where not a soul knew her. She took a year's leave, and went to Melbourne.

People were kind. At dinner parties in terrace houses they said to her, Of course Queensland gets the kind of government it deserves do you like the linguini? the salmon? in Brisbane we thought the food perfectly *ghastly* we do congratulate you on leaving, oh the Queensland police, the Aboriginal problem, no awareness at all, and Namatjira's tonal effects are *exquisite,* there was a black tie opening and we were simply overwhelmed *overwhelmed,* Aboriginal art's the going thing now a fantastic investment and I myself have a poem of social protest, a very *meaningful* people were kind enough a very socially aware in the *Age* will you have more champagne? you've come out of the wilderness, they said.

She was mute. The same hollow alphabet. No. Hollower. She could not acquire the knack of words that floated so weightlessly. She fled back to Queensland. She dreamed of alphabets that sent down deep webbing roots.

At dawn on the Burkedin banks (is she sixty or twenty?) she watches the foraging parties. Swimmers, dinghies, fights, whoops of delight. She huddles, not wanting to be found. Whole doors are coming up, chairs, verandah spindles, stovepipes, crosses, bits of clapboard, signs, signposts, there's a black market trade in souvenirs. At a trestle table, t-shirts are selling like hot cakes: *I was there for the Second Coming of Come-by-Chance.*

Adeline sits hugging herself, shivering in the fierce morning heat. Two gangs are fighting over the clock face from the Post Office tower, and a reporter in a frenzy of picture-taking swears irritably as he runs out of film. Rewind, unload, *rip* (the velcro carry-case), *rip* (the kodak pack), *rip* (the foil covering). "Shit!" He tosses packet and foil over his shoulder. He is not an ordinary reporter. He has literary sensibilities and does these things, these projects, as a cultural enterprise, a refined monitoring of the pulse of the nation. He shakes his head at Adeline in disbelief. "I don't believe this. Bloody animals, a pack of hooligan looters." He gives off a kind of jubilation of disgust. "No one's going to *believe* this in Melbourne."

"No. No one ever believes." Nevertheless, that does not absolve .... She takes a deep breath. "I would like to set the record straight."

"Yeah, who wouldn't?" He's got the clock face and a bloodied forehead in focus, he's shooting like crazy.

"Capper is not my real name, I was never married, I am Adeline Crick."

"Yeah?"

"I want to tell you what really happened."

Christ, not another one. Any direction you point your lens. And she's got the DTs, the old soak, she's only worth one shot.

Adeline's words are heavy, their roots go down below the Burdekin, her clumsy tongue trips on them, she has to speak with the care of those who have had a stroke. She says: "I have the blood of innocent men on my hands."

"Oh," he says. "Right."

"At times one has to ask oneself," he wrote in a photoessay that was given prominent space in the *Age,* "if Queensland is our own Gothic invention, a kind of morality play, the Bosch canvas of the Australian psyche, a sort of perpetual *memento mori* that points to the frailty of the skein of civilization reaching out so tentatively from our southern cities.

"To return to Sydney or Melbourne and write of the primitive violence, the yobbo mentality, the mystics, the pathetic old women generating lurid and gratuitous confessions, the general sense of mass hallucination ... to speak of this is to risk charges of sensationalism. And indeed, after mere days back in the real world, one has the sense of emerging from a drugged and aberrant condition.

"One has to ask oneself: Does Queensland actually exist?

"And one has to conclude: I think not.

"Queensland is a primitive state of mind from which the great majority of us, mercifully, have long since evolved. And Come-by-Chance is a dream within a nightmare, the hysteric's utopia, the city of Robespierre, Stalin, Jim Jones, the vision of purity from which history recoils.

"Come-by-Chance, we who are sane dilute you."

Yes, he'd done that rather well. Seen the essence of things, touched the depths, but kept the tone right. Words were his business, and if he often caught himself being plangent and acute, well, it was a forgivable sin. He was tempted to add a rider explaining how his work should be read, how his words should be picked up one by one like stones from the bank of an enchanted creek. But he would save that for another time.

When the drought broke with the series of maverick cyclones we all remember, there was flash flooding throughout central and southern Queensland. At the tent city on the Burdekin Dam, winds hurling themselves down from the Gulf at unprecedented inland speeds caused death and mayhem. Police estimated as many as forty people drowned. Cars were marooned on the Flinders Highway for days, army ducks were still rescuing stranded survivors weeks later. In both coastal and inland cities, powerline disasters, the uprooting of trees, and the collapse of buildings in the gale-force winds brought the region's death toll to over one hundred.

In Melbourne and Sydney, where water restrictions were at last lifted to everyone's immense relief, people read of the Queensland floods and shook their heads. If it's not one thing, it's another, they said.

A FAMILY Supper

KAZUO ISHIGURO

FUGU IS A FISH CAUGHT OFF THE PACIFIC shores of Japan. The fish has held a special significance for me ever since my mother died through eating one. The poison resides in the sexual glands of the fish, inside two fragile bags. When preparing the fish, these bags must be removed with caution, for any clumsiness will result in the poison leaking into the veins. Regrettably, it is not easy to tell whether or not this operation has been carried out successfully. The proof is, as it were, in the eating.

Fugu poisoning is hideously painful and almost always fatal. If the fish has been eaten during the evening, the victim is usually overtaken by pain during his sleep. He rolls about in agony for a few hours and is dead by morning. The fish became extremely popular in Japan after the war. Until stricter regulations were imposed, it was all the rage to perform the hazardous gutting operation in one's own kitchen, then to invite neighbours and friends round for the feast.

At the time of my mother's death, I was living in California. My relationship with my parents had become somewhat strained around that period, and consequently I did not learn of the circumstances surrounding her death until I returned to Tokyo two years later. Apparently, my mother had always refused to eat fugu, but on this particular occasion she had made an exception, having been invited by an old schoolfriend whom she was anxious not to offend. It was my father who supplied me with the details as we drove from the airport to his house in the Kamakura district. When

Kazuo Ishiguro was born in Nagasaki, Japan in 1954 and moved to Britain when he was five: "But as a child, I grew thinking I was going to return to Japan any day. And so I had this very powerfully imagined country in my head... (Later, as a writer) I wasn't interested in doing research in the conventional sense to fill out my picture of Japan. I had a Japan inside my head, which I needed to transcribe as accurately as possible." Ishiguro won the Booker Prize in 1989.

we finally arrived, it was nearing the end of a sunny autumn day.

"Did you eat on the plane?" my father asked. We were sitting on the tatami floor of his tea-room.

"They gave me a light snack."

"You must be hungry. We'll eat as soon as Kikuko arrives."

My father was a formidable-looking man with a large stony jaw and furious black eyebrows. I think now in retrospect that he much resembled Chou En-lai, although he would not have cherished such a comparison, being particularly proud of the pure samurai blood that ran in the family. His general presence was not one which encouraged relaxed conversation; neither were things helped much by his odd way of stating each remark as if it were the concluding one. In fact, as I sat opposite him that afternoon, a boyhood memory came back to me of the time he had struck me several times around the head for "chattering like an old woman." Inevitably, our conversation since my arrival at the airport had been punctuated by long pauses.

"I'm sorry to hear about the firm," I said when neither of us had spoken for some time. He nodded gravely.

"In fact the story didn't end there," he said. "After the firm's collapse, Watanabe killed himself. He didn't wish to live with the disgrace."

"I see."

"We were partners for seventeen years. A man of principle and honour. I respected him very much."

"Will you go into business again?" I asked.

"I am — in retirement. I'm too old to involve myself in new ventures now. Business these days has become so different. Dealing with foreigners. Doing things their way. I don't understand how we've come to this. Neither did Watanabe." He sighed. "A fine man. A man of principle."

The tea-room looked out over the garden. From where I sat I could make out the ancient well which as a child I had believed haunted. It was just visible now through the thick foliage. The sun had sunk low and much of the garden had fallen into shadow.

"I'm glad in any case that you've decided to come back," my father said. "More than a short visit, I hope."

"I'm not sure what my plans will be."

"I for one am prepared to forget the past. Your mother too was always ready to welcome you back — upset as she was by your behaviour."

"I appreciate your sympathy. As I say, I'm not sure what my plans are."

"I've come to believe now that there were no evil intentions in your mind," my father continued. "You were swayed by certain — influences. Like so many others."

"Perhaps we should forget it, as you suggest."

"As you will. More tea?"

Just then a girl's voice came echoing through the house.

"At last." My father rose to his feet. "Kikuko has arrived."

Despite our difference in years, my sister and I had always been close. Seeing me again seemed to make her excessively excited and for a while she did nothing but giggle nervously. But she calmed down somewhat when my father started to question her about Osaka and her university. She answered him with short formal replies. She in turn asked me a few questions, but she seemed inhibited by the fear that her questions might lead to awkward topics. After a while, the conversation had become even sparser than prior to Kikuko's arrival. Then my father stood up, saying: "I must attend to the supper. Please excuse me for being burdened down by such matters. Kikuko will look after you."

My sister relaxed quite visibly after he had left the room. Within a few minutes, she was chatting freely about her friends in Osaka and about her classes at university. Then quite suddenly she decided we should walk in the garden and went striding out onto the veranda. We put on some straw sandals that had been left along the veranda rail and stepped out into the garden. The daylight had almost gone.

"I've been dying for a smoke for the last half-hour," she said, lighting a cigarette.

"Then why didn't you smoke?"

She made a furtive gesture towards the house then grinned mischievously.

"Oh I see," I said.

"Guess what? I've got a boyfriend now."

"Oh yes?"

"Except I'm wondering what to do. I haven't made up my mind yet."

"Quite understandable."

"You see, he's making plans to go to America. He wants me to go with him as soon as I finish studying."

"I see. And you want to go to America?"

"If we go, we're going to hitchhike." Kikuko waved a thumb in front of my face. "People say its dangerous, but I've done it in Osaka and it's fine."

"I see. So what is it you're unsure about?"

We were following a narrow path that wound through the shrubs and finished by the old well. As we walked, Kikuko persisted in taking unnecessarily theatrical puffs on her cigarette.

"Well. I've got lots of friends now in Osaka. I like it there. I'm not sure I want to leave them all behind just yet. And Suichi — I like him, but I'm not sure I want to spend so much time with him. Do you understand?"

"Oh perfectly."

She grinned again, then skipped on ahead of me until she had reached the well. "Do you remember," she said, as I came walking up to her, "how you used to say this well was haunted?"

"Yes, I remember."

We both peered over the side.

"Mother always told me it was the old woman from the vegetable store you'd seen that night," she said. "But I never believed her and never came out here alone."

"Mother used to tell me that too. She even told me once the old woman had confessed to being the ghost. Apparently she'd been taking a short cut through our garden. I imagine she had some trouble clambering over these walls."

Kikuko gave a giggle. She then turned her back to the well, casting her gaze about the garden.

"Mother never really blamed you, you know," she said, in a new voice. I remained silent. "She always used to say to me how it was their fault, her's and Father's, for not bringing you up correctly. She used to tell me how much more careful they'd been with me, and that's why I was so good." She looked up and the mischievous grin had returned to her face. "Poor Mother," she said.

"Yes. Poor Mother."

"Are you going back to California?"

"I don't know. I'll have to see."

"What happened to — to her? To Vicki?"

"That's all finished with," I said. "There's nothing much left for me now in California."

"Do you think I ought to go there?"

"Why not? I don't know. You'll probably like it." I glanced towards the house. "Perhaps we'd better go in soon. Father might need a hand with supper."

But my sister was once more peering down into the well. "I can't see any ghosts," she said. Her voice echoed a little.

"Is Father very upset about his firm collapsing?"

"Don't know. You can never tell with Father." Then suddenly she straightened up and turned to me. "Did he tell you about Watanabe? What he did?"

"I heard he committed suicide."

"Well, that wasn't all. He took his whole family with him. His wife and his two little girls."

"Oh yes?"

"Those two beautiful little girls. He turned on the gas while they were all asleep. Then he cut his stomach with a meat knife."

"Yes, Father was just telling me how Watanabe was a man of principle."

"Sick." My sister turned back to the well.

"Careful. You'll fall right in."

"I can't see any ghost," she said. "You were lying to me all that time."

"But I never said it lived down the well."

"Where is it then?"

We both looked around at the trees and shrubs. The light in the garden had grown very dim. Eventually I pointed to a small clearing some ten yards away.

"Just there I saw it. Just there."

We stared at the spot.

"What did it look like?"

"I couldn't see very well. It was dark."

"But you must have seen something."

"It was an old woman. She was just standing there, watching me."

We kept staring at the spot as if mesmerized.

"She was wearing a white kimono," I said. "Some of her hair had come undone. It was blowing around a little."

Kikuko pushed her elbow against my arm. "Oh be quiet. You're trying to frighten me all over again." She trod on the remains of her cigarette, then for a brief moment stood regarding it with a perplexed expression. She kicked some pine needles over it, then once more displayed her grin. "Let's see if supper's ready," she said.

We found my father in the kitchen. He gave us a quick glance, then carried on with what he was doing.

"Father's become quite a chef since he's had to manage on his own," Kikuko said with a laugh. He turned and looked at my sister coldly.

"Hardly a skill I'm proud of," he said. "Kikuko, come here and help."

For some moments my sister did not move. Then she stepped forward and took an apron hanging from a drawer.

"Just these vegetables need cooking now," he said to her. "The rest just needs watching." Then he looked up and regarded me strangely for some seconds. "I expect you want to look around the house," he said eventually. He put down the chopsticks he had been holding. "It's a long time since you've seen it."

As we left the kitchen I glanced back towards Kikuko, but her back was turned.

"She's a good girl," my father said quietly.

I followed my father from room to room. I had forgotten how large the house was. A panel would slide open and another room would appear. But the rooms were all startlingly empty. In one of the rooms the lights did not come on, and we stared at the stark walls and tatami in the pale light that came from the windows.

"This house is too large for a man to live in alone," my father said. "I don't have much use for most of these rooms now."

But eventually my father opened the door to a room packed full of books and papers. There were flowers in vases and pictures on the walls. Then I noticed something on a low table in the corner of the room. I came nearer and saw it was a plastic model of a battleship, the kind constructed by children. It had been placed on some newspaper; scattered around it were assorted pieces of grey plastic.

My father gave a laugh. He came up to the table and picked up the model.

"Since the firm folded," he said, "I have a little more time on my hands." He laughed again, rather strangely. For a moment his face looked almost gentle. "A little more time."

"That seems odd," I said. "You were always so busy."

"Too busy perhaps." He looked at me with a small smile. "Perhaps I should have been a more attentive father."

I laughed. He went on contemplating his battleship. Then he looked up. "I hadn't meant to tell you this, but perhaps it's best that I do. It's my belief that your mother's death was no accident. She had many worries. And some disappointments."

We both gazed at the battleship.

"Surely," I said eventually, "my mother didn't expect me to live here for ever."

"Obviously you don't see. You don't see how it is for some parents. Not only must they lose their children, they must lose them to things they don't understand." He spun the battleship in his fingers. "These little gunboats here could have been better glued, don't you think?"

"Perhaps. It looks fine."

"During the war I spent some time on a ship rather like this. But my ambition was always the air force. I figured it like this. If your ship was struck by the enemy, all you could do was struggle in the water hoping for a lifeline. But in the airplane — well — there was always the final weapon." He put the model back on the table. "I don't suppose you believe in war."

"Not particularly."

He cast an eye around the room. "Supper should be ready by now," he said. "You must be hungry."

Supper was waiting in a dimly lit room next to the kitchen. The only source of light was a big lantern that hung over the table, casting the rest of the room into shadow. We bowed to each other before starting the meal.

There was little conversation. When I made a polite

comment about the food, Kikuko giggled a little. Her earlier nervousness seemed to have returned to her. My father did not speak for several minutes. Finally he said:

"It must feel strange for you, being back in Japan."

"Yes, it is a little strange."

"Already, perhaps, you regret leaving America."

"A little. Not so much. I didn't leave behind much. Just some empty rooms."

"I see."

I glanced across the table. My father's face looked stony and forbidding in the half-light. We ate on in silence.

Then my eye caught something at the back of the room. At first I continued eating, then my hands became still. The others noticed and looked at me. I went on gazing into the darkness past my father's shoulder.

"Who is that? In that photograph there?"

"Which photograph?" My father turned slightly, trying to follow my gaze.

"The lowest one. The old woman in the white kimono."

My father put down his chopsticks. He looked at the photograph, then at me.

"Your mother." His voice had become very hard. "Can't you recognize your own mother?"

"My mother. You see, it's dark. I can't see it very well."

No one spoke for a few seconds, then Kikuko rose to her feet. She took the photograph off the wall, then came back to the table and gave it to me.

"She looks a lot older," I said.

"It was taken shortly before her death," said my father.

"It was the dark. I couldn't see it very well."

I looked up and noticed my father holding out a hand. I gave him the photograph. He looked at it intently, then held it towards Kikuko. Obediently, my sister rose to her feet once more and returned the picture to the wall.

There was a large pot left unopened at the centre of the table. When Kikuko had seated herself again, my father reached forward and lifted the lid. A cloud of steam rose up and curled towards the lantern. He pushed the pot a little towards me.

"You must be hungry," he said. One side of his face had fallen into shadow.

"Thank you." I reached forward with my chopsticks. The steam was almost scalding. "What is it?"

"Fish."

"It smells very good."

In amidst the soup were strips of fish that had curled almost into balls. I picked one out and brought it to my bowl.

"Help yourself. There's plenty."

"Thank you." I took a little more, then pushed the pot towards my father. I watched him take several pieces to his bowl then we both watched as Kikuko served herself.

My father bowed slightly. "You must be hungry," he said again. He took some fish to his mouth and started to eat. Then I too chose a piece and put it in my mouth. It felt quite soft, quite fleshy against my tongue.

"Very good," I said, "What is it?"

"Just fish."

"It's very good."

The three of us ate on in silence. Several minutes went by.

"Some more?"

"Is there enough?"

"There's plenty for all of us." My father lifted the lid and once more steam rose up. We all reached forward and helped ourselves.

"Here," I said to my father, "you have this last piece."

"Thank you."

When we finished the meal, my father stretched out his arms and yawned with an air of satisfaction. "Kikuko," he said. "Prepare a pot of tea, please."

My sister looked at him, then left the room without comment. My father stood up.

"Let's retire to the other room. It's rather warm in here."

I got to my feet and followed him into the tea-room. The large sliding windows had been left open, bringing in a breeze from the garden. For a while we sat in silence.

"Father," I said, finally.

"Yes?"

"Kikuko tells me Watanabe-San took his whole family with him."

My father lowered his eyes and nodded. For some moments he seemed in deep thought. "Watanabe was very devoted to his work," he said at last. "The collapse of the firm was a great blow to him. I fear it must have weakened his judgement."

"You think what he did — it was a mistake?"

"Why, of course. Do you see it otherwise?"

"No, no. Of course not."

"There are other things besides work."

"Yes."

We fell silent again. The sounds of the locusts came in from the garden. I looked out into the darkness. The well was no longer visible.

"What do you think you will do now?" my father asked. "Will you stay in Japan for a while?"

"To be honest, I hadn't thought that far ahead."

"If you wish to stay here, I mean this house, you would be very welcome. That is, if you don't mind living with an old man."

"Thank you. I'll have to think about it."

I gazed out once more into the darkness.

"But of course," said my father, "this house is so dreary now. You'll no doubt be returning to America before long."

"Perhaps. I don't know yet."

"No doubt you will."

For some time my father seemed to be studying the back of his hands. Then he looked up and sighed.

"Kikuko is due to complete her studies next spring," he said. "Perhaps she will want to come home then. She's a good girl."

"Perhaps she will."

"Things will improve then."

"Yes, I'm sure they will."

We fell silent once more, waiting for Kikuko to bring the tea.

The
BURNING
Roadblocks

RYSZARD KAPUSCINSKI

JANUARY 1966. IN NIGERIA A CIVIL WAR was going on. I was a correspondent covering the war. On a cloudy day I left Lagos. On the outskirts police were stopping all cars. They were searching all the trunks, looking for weapons. They ripped open sacks of corn: could there be ammunition in that corn?

Authority ended at the city line.

The road leads through a green countryside of low hills covered with a close, thick bush. This is a laterite road, rust-coloured, with a treacherous uneven surface.

Ryszard Kapuscinski was born in 1932 in the city of Pinsk in eastern Poland. Working for twenty-two years as a foreign correspondent for the Polish Press Agency, he has covered twenty-seven revolutions and coups. Again and again, he conjures a personal landscape of dislocation: of finding oneself over and over again "alone in somebody else's country during somebody else's war."

These hills, this road and the villages along it are the country of the Yorubas, who inhabit south-western Nigeria. They constitute a quarter of Nigeria's population. The heaven of the Yorubas is full of gods and their earth full of kings. The greatest god is called Oduduwa and he lives at a height higher than the stars, higher even than the sun. The kings, on the other hand, live close to the people. In every city and every village there is a king.

In 1962 the Yorubas split into two camps. The overwhelming majority belongs to the UPGA (United Progressive Grand Alliance); an insignificant minority belongs to the NNDP (Nigerian National Democratic Party). Owing to the trickery of the Nigerian central government, the minority party rules the Yorubas' province. The central government prefers a minority government in the province as a way of controlling the Yorubas and curbing their separatist ambitions: thus has the party of the overwhelming majority, the UPGA, found itself in opposition. It was obvious that the majority party, UPGA, had won. Nevertheless, the central government, ignoring the results and the sentiments of the Yorubas, declared the victory of the puppet NNDP, which went on to form a government. In protest, the majority party created its own government. For a time there were two governments. In the end, the members of the majority government were imprisoned, and the UPGA declared open war against the minority government.

And so we have misfortune, we have a war. It is an unjust, dirty, hooliganish war in which all methods are allowed — whatever it takes to knock out the opponent and gain control. This war uses a lot of fire: houses are burning, plantations are burning, and charred bodies lie in the streets.

The whole land of the Yorubas is in flames.

I was driving along a road where they say no white man can come back alive. I was driving to see if a white man could because I had to experience everything for myself. I know that a man shudders in the forest when he passes close to a lion. I got close to a lion so that I would know how it feels. I had to do it myself because I knew no one could describe it to me. And I cannot describe it myself. Nor can I describe a night in the Sahara. The stars over the Sahara are enormous. They sway above the sand like great chandeliers. The light of those stars is green. Night in the Sahara is as green as a Mazowsze meadow.

I might see the Sahara again and I might see the road that carried me through Yoruba country again. I drove up a hill and when I got to the crest I could see the first flaming roadblock down below.

It was too late to turn back.

Burning logs blocked the road. There was a big bonfire in the middle. I slowed down and then stopped; it would have been impossible to have carried on. I could see a dozen or so young people. Some had shotguns, some were holding knives and the rest were armed with machetes. They were dressed alike in blue shirts with white sleeves, the colours of the opposition, of the UPGA. They wore black and white caps with the letters UPGA. They had pictures of Chief Awolowo pinned to their shirts. Chief Awolowo was the leader of the opposition, the idol of the party.

I was in the land of UPGA activists. They must have been smoking hashish because their eyes were mad and they did not look fully conscious. They were soaked in sweat, seemed possessed, frenzied.

They descended on me and pulled me out of the car. I could hear them shouting "UPGA! UPGA!" On this road, UPGA ruled. UPGA held me in its sway. I could feel three knife-points against my back and I saw several machetes (these are the African scythes) aimed at my head. Two activists stood a few steps away, pointing their guns at me in case I tried to get away. I was surrounded. Around me I could see sweaty faces with jumpy glances: I could see knives and gun barrels.

My African experience had taught me that the worst thing to do in such situations is to betray your despair; the worst thing is to make a gesture of self-defence, because that emboldens them, because that unleashes a new wave of aggression in them.

In the Congo when they poked machine-guns in our bellies, we could not flinch. The most important thing was keeping still. Keeping still takes practice and willpower, because everything inside screams that you should run for it or jump the other guy. But they are always in groups and that means certain death. This was a moment when he, the black, was testing me, looking for a weak spot. He would have been afraid of attacking my strong point — he had too much fear of the white in him — so he looked for my weakness. I had to cover all my weaknesses, hide them somewhere very deep within myself. This was Africa, I was in Africa. They did not know that I was not their enemy. They knew that I was white, and the only white they had known was the colonizer, who abased them, and now they wanted to make him pay for it.

The irony of the situation was that I would die out of responsibility for colonialism; I would die in expiation of the slave merchants: I would die to atone for the white planter's whip; I would die because Lady Lugard had ordered them to carry her in her litter.

The ones standing in the road wanted cash. They wanted me to join the party, to become a member of UPGA and to pay for it. I gave them five shillings. That was too little, because somebody hit me on the back of the head. I felt pain in my skull. In a moment there was another blow. After the third blow I felt an enormous tiredness. I was fatigued and sleepy; I asked how much they wanted.

They wanted five pounds.

Everything in Africa was getting more expensive. In the

Congo soldiers were accepting people into the party for one pack of cigarettes and one blow with a rifle butt. But here I had already got it a couple of times and I was still supposed to pay five pounds. I must have hesitated because the boss shouted to the activists, "Burn the car!" and that car, the Peugeot that had been carrying me around Africa, was not mine. It belonged to the Polish state. One of them splashed gasoline on to the Peugeot.

I understood that the discussion had ended and I had no way out. I gave them the five pounds. They started fighting over it.

But they allowed me to drive on. Two boys moved the burning logs aside. I looked around. On both sides of the road there was a village and the village crowd had been watching the action. The people were silent; somebody in the crowd was holding up a UPGA banner. They all had photographs of Chief Awolowo pinned on their shirts. I liked the girls best. They were naked to the waist and had the name of the party written across their breasts: UP on the right breast, and GA on the left one.

I started off.

I could not turn back; they allowed me only to go forward. So I kept driving through a country at war, a cloud of dust behind me. The landscape was beautiful here, all vivid colours, Africa the way I like it. Quiet, empty — every now and then a bird taking flight in the path of the car. The roaring of a factory was only in my head. But an empty road and a car gradually restore calm.

Now I knew the price: UPGA had demanded five pounds of me. I had less than five pounds left, and fifty kilometres to go. I passed a burning village and then an emptying village, people fleeing into the bush. Two goats grazed by the roadside and smoke hung above the road.

Beyond the village there was another burning roadblock.

Activists in UPGA uniforms, knives in their hands, were kicking a driver who did not want to pay his membership fee. Nearby stood a bloody, beaten man — he hadn't been able to come up with the dues, either. Everything looked like the first roadblock. At this one, though, I hadn't even managed to announce my desire to join UPGA before I received a pair of hooks to the midsection and had my shirt torn. They turned my pockets inside out and took all my money.

I was waiting for them to set me on fire, because UPGA was burning many people alive. I had seen the burnt corpses. The boss at this roadblock popped me in the face and I felt a warm sweetness in my mouth. Then he poured benzene on me, because here they burn people in benzene: it guarantees complete incineration.

I felt an animal fear, a fear that struck me with paralysis: I stood rooted to the ground, as if I was buried up to the neck. I could feel the sweat flowing over me, but under my skin I was as cold as if standing naked in sub-zero frost.

I wanted to live but life was abandoning me. I wanted to live, but I did not know how to defend my life. My life was going to end in inhuman torment. My life was going to go out in flames.

What did they want from me? They waved a knife before my eyes. They pointed it at my heart. The boss of the operation stuffed my money into his pocket and shouted at me, blasting me with his beery breath: "Power! UPGA must get power! We want power! UPGA is power!" He was shaking, swept up in the passion of power; he was mad on power; the very word "power" sent him into ecstasy, into the highest rapture. His face was flooding sweat, the veins on his forehead were bulging and his eyes were shot with blood and madness. He was happy and he began to laugh in joy. They all started laughing. That laughter saved me.

They ordered me to drive on.

The little crowd around the roadblock shouted "UPGA!" and held up their hands with two fingers stretched out in the "V" sign: Victory for UPGA on all fronts.

About four kilometres down the road the third roadblock was burning. The road was straight and I could see the smoke a long way off, and then I saw the fire and the activists. I could not turn back. There were two barriers behind me. I could only go forward. I was trapped, falling out of one ambush and into another. But now I was out of money for ransom, and I knew that if I didn't pay up they would burn my car. Above all, I didn't want another beating. I had been whipped, my shirt was in tatters and I reeked of benzene.

There was only one way out: to run the roadblock. It was risky, because I might wreck the car or it might catch fire. But I had no choice.

I floored it. The roadblock was a kilometre ahead. The

speedometer needle jumped: 110, 120, 140. The car shimmied and I gripped the wheel more tightly. I leaned on the horn. When I was right on top of it I could see the bonfire stretched all the way across the road. The activists were waving their knives for me to stop. I saw that two of them were winding up to throw bottles of gasoline at the car and for a second I thought, so, this is the end, but there was no turning back. There was no turning ...

I smashed into the fire, the car jumped, there was a hammering against the belly pan, sparks showered over the windshield. And suddenly — the roadblock, the fire and the shouting were behind me. The bottles had missed. Hounded by terror, I drove another kilometre and then I stopped to make sure the car wasn't on fire. It wasn't on fire. I was all wet. All my strength had left me; I was incapable of fighting; I was wide open, defenceless. I sat down on the sand and felt sick to my stomach. Everything around me was alien. An alien sky and alien trees. Alien hills and manioc fields. I couldn't stay there, so I got back in and drove until I came to a town called Idiroko. On the way I passed a police station and I stopped there. The policemen were sitting on a bench. They let me wash and straighten myself out.

I wanted to return to Lagos, but I couldn't go back alone. The commandant started to organize an escort. But the policemen were afraid to travel alone. They needed to borrow a car, so the commandant went into town. I sat on a bench reading the *Nigerian Tribune*, the UPGA paper. The paper was dedicated to party activities and the party's fight for power. "Our furious battle," I read, "is continuing. For instance, our activists burned the eight-year old pupil Janet Bosede Ojo of Ikerre alive. The girl's father had voted for the NNDP." I read on: "In Ilesha the farmer Alek Aleke was burned alive. A group of activists used the 'Spray-and-Lite' method (also known as 'UPGA candles') on him. The farmer was returning to his fields when the activists grabbed him and commanded him to strip naked. The farmer undressed, fell to his knees and begged for mercy. In this position he was sprayed with benzene and set afire." The paper was full of similar reports. UPGA was fighting for power, and the flames of that struggle were devouring people.

The commandant returned, but without a car. He designated three policemen to ride in mine. They were afraid to go. In the end they got in, pointed their rifles out the windows, and we drove off that way, as if in an armoured vehicle. At the first roadblock the fire was still burning but there was nobody in sight. The next two roadblocks were in full swing, but when they saw the police they let us through. The policemen weren't going to allow the car to be stopped; they didn't want to get in a fight with the activists. I understood — they lived here and wanted to survive. Today they had rifles, but usually they went unarmed. Many policemen had been killed in the region.

At dusk we were in Lagos.

A SMALL Place

JAMAICA KINCAID

Jamaica Kincaid was born in St. John's, Antigua in 1949. She describes the reader's dilemma of wishing to lay claim to an invented country in a borrowed book: "When I was growing up ... I stole many books from {the} library. I didn't mean to ...; it's just that once I had read a book I couldn't bear to part with it."

IF YOU GO TO ANTIGUA AS A TOURIST, THIS IS what you will see. If you come by airplane, you will land at the V.C. Bird International Airport. Vere Cornwall (V.C.) Bird is the Prime Minister of Antigua. You may be the sort of tourist who would wonder why a Prime Minister would want an airport named after him — why not a school, why not a hospital, why not some great public monument? You are a tourist and you have not yet seen a school in Antigua, you have not yet seen the hospital in Antigua, you have not yet seen a public monument in Antigua. As your plane descends to land, you might say, What a beautiful island Antigua is — more beautiful than any of the other islands you have seen, and they were very beautiful, in their way, but they were much too green, much too lush with vegetation, which indicated to you, the tourist, that they got quite a bit of rainfall, and rain is the very thing that you, just now, do not want, for you are thinking of the hard and cold and dark and long days you spent working in North America (or, worse, Europe), earning some money so that you could stay in this place (Antigua) where the sun always shines and where the climate is deliciously hot and dry for the four to ten days you are going to be staying there; and since you are on your holiday, since you are a tourist, the thought of what it might be like for someone who had to live day in, day out in a place that suffers constantly from drought, and so has to watch carefully every drop of fresh water used (while at the same time surrounded by a sea and an ocean — the Caribbean Sea on one side, the Atlantic Ocean on the other), must never cross your mind.

You disembark from your plane. You go through customs.

Since you are a tourist, a North American or European — to be frank, white — and not an Antiguan black returning to Antigua from Europe or North America with cardboard boxes of much needed cheap clothes and food for relatives, you move through customs swiftly, you move through customs with ease. Your bags are not searched. You emerge from customs into the hot, clean air: immediately you feel cleansed, immediately you feel blessed (which is to say special); you feel free. You see a man, a taxi driver; you ask him to take you to your destination; he quotes you a price. You immediately think that the price is in the local currency, for you are a tourist and you are familiar with these things (rates of exchange) and you feel even more free, for things seem so cheap, but then your driver ends by saying, "In U.S. currency." You may say, "Hmmmm, do you have a formal sheet that lists official prices and destinations?" Your driver obeys the law and shows you the sheet, and he apologizes for the incredible mistake he has made in quoting you a price off the top of his head which is so vastly different (favouring him) from the one listed. You are driven to your hotel by this taxi driver in his taxi, a brand new Japanese-made vehicle. The road on which you are travelling is a very bad road, very much in need of repair. You are feeling wonderful, so you say, "Oh, what a marvellous change these bad roads are from the splendid highways I am used to in North America." (Or, worse, Europe.) Your driver is reckless; he is a dangerous man who drives in the middle of the road when he thinks no other cars are coming in the opposite direction, passes other cars on blind curves that run uphill, drives at sixty miles an hour on narrow, curving roads when the road sign, a rusting, beat-up thing left over from colonial days, says 40 MPH. This might frighten you (you are on your holiday; you are a tourist); this might excite you (you are on your holiday; you are a tourist), though if you are from New York and take taxis you are used to this style of driving: most of the taxi drivers in New York are from places in the world like this. You are looking out the window (because you want to get your money's worth); you notice that all the cars you see are brand-new, or almost brand-new, and that they are all Japanese-made. There are no American cars in Antigua — no new ones, at any rate; none that were manufactured in the last ten years. You continue to look at the cars and you say to yourself, Why, they look brand-new, but they have an awful sound, like an old car — a very old, dilapidated car. How to account for that? Well, possibly it's because they use leaded gasoline in these brand-new cars whose engines were built to use non-leaded gasoline, but you musn't ask the person driving the car if this is so, because he or she has never heard of unleaded gasoline. You look closely at the car; you see that it's a model of a Japanese car that you might hesitate to buy; it's a model that's very expensive; it's a model that's quite impractical for a person who has to work as hard as you do and who watches every penny you earn so that you can afford this holiday you are on. How do they afford such a car? And do they live in a luxurious house to match such a car? Well, no. You will be surprised, then, to see that most likely the person driving this brand new car filled with the wrong gas lives in a house that, in comparison, is far beneath the status of the car; and if you were to ask why you would be told that the banks are encouraged by the government to make loans available for cars, but loans for houses not so easily available; and if you ask again why, you will be told that the two main car dealerships in Antigua are owned in part or outright by ministers in government. Oh, but you are on holiday and the sight of these brand-new cars driven by people who may or may not have really passed their driving test (there was once a scandal about driving licenses for sale) would not really stir up these thoughts in you. You pass a building sitting in a sea of dust and you think, It's some latrines for people just passing by, but when you look again you see the building has written on it PIGGOT'S SCHOOL. You pass the hospital, the Holberton Hospital, and how wrong you are not to think about this, for though you are a tourist on your holiday, what if your heart should miss a few beats? What if a blood vessel in your neck should break? What if one of those people driving those brand-new cars filled with the wrong gas fails to pass safely while going uphill on a curve and you are in the car going in the opposite direction? Will you be comforted to know that the hospital is staffed with doctors that no actual Antiguan trusts; that Antiguans always say about the doctors, "I don't want them near me"; that Antiguans refer to them not as doctors but as "the three men" (there are three of them); that when the Minister of Health himself doesn't feel well he takes the first plane to New York to see a real doctor; that if any one of the ministers in government needs medical care he flies to New York to get it?

It's a good thing that you brought your own books with you, for you couldn't just go to the library and borrow some. Antigua used to have a splendid library, but in The Earthquake (everyone talks about it that way — The Earthquake; we Antiguans, for I am one, have a great sense of things, and the more meaningful the thing, the more meaningless we make it) the library building was damaged. This was in 1974, and soon after that a sign was placed on the front of the building saying, THIS BUILDING WAS DAMAGED IN THE EARTHQUAKE OF 1974. REPAIRS ARE PENDING. The sign hangs there, and hangs there more than a decade later, with its unfulfilled promise of repair, and you might see this as a sort of quaintness on the part of these islanders, these people descended from slaves — what a strange, unusual perception of time they have. REPAIRS ARE PENDING, and here it is, many years later, but perhaps in a world that is twelve miles long and nine miles wide (the size of Antigua) twelve minutes and twelve days are all the same. The library is one of those splendid old buildings from colonial times, and the sign telling of the repairs is a splendid old sign from colonial times. Not long after The Earthquake Antigua got its independence from Britain, making Antigua a state in its own right, and Antiguans are so proud of this that each year, to mark the day, they go to Church and thank God, a British God, for this. But you should not think of the confusion that must lie in all that and you must not think of the damaged library. You have brought your own books with you, and among them is one of those new books about economic history, one of those books explaining how the West (meaning Europe and North America after its conquest and settlement by Europeans) got rich: the West got rich not from free (free — in this case, meaning got-for-nothing) and then undervalued labour, for generations, of the people like me you see walking around you in Antigua but from the ingenuity of small shopkeepers in Sheffield and Yorkshire and Lancashire, or wherever; and what a part the invention of the wristwatch played in it, for there was nothing noble-minded men could not do when they discovered they could slap time on their wrists just like that (isn't that the last straw; for not only did we have to suffer the unspeakableness of slavery, but the satisfaction to be had from "We made you bastards rich" is taken away, too), and so you needn't let that slightly funny feeling you have from time to time about exploitation, oppression, domination develop into full-fledged unease, discomfort; you could ruin your holiday. They are not responsible for what you have; you owe them nothing; in fact, you did them a big favour, and you can provide one hundred examples. For here you are now, passing by Government House. And here you are now, passing by the Prime Minister's Office and the Parliament Building, and overlooking these, with a splendid view of St. John's Harbour, the American Embassy. If it were not for you, they would not have the Government House, and Prime Minister's Office, and Parliament Building and embassy of a powerful country. Now you are passing a mansion, an extraordinary house painted the colour of cow dung, with more aerials and antennas attached to it than you will see even at the American Embassy. The people who live in this house are a merchant family who came to Antigua from the Middle East less than twenty years ago. When this family first came to Antigua they sold dry goods door to door from suitcases they carried on their backs. Now they own a lot of Antigua; they regularly lend money to the government, they build enormous (for Antigua), ugly (for Antigua), concrete buildings in Antigua's capital, St. John's, which the government then rents for huge sums of money; a member of their family is the Antiguan Ambassador to Syria; Antiguans hate them. Not far from this mansion is another mansion, the home of a drug smuggler. Everybody knows he's a drug smuggler, and if just as you were driving by he stepped out of his door your driver might point him out to you as the notorious person that he is, for this drug smuggler is so rich people say he buys cars in tens — ten of this one, ten of that one — and that he bought a house (another mansion) near Five Islands, contents included, with cash he carried in a suitcase: three hundred and fifty thousand American dollars, and, to the surprise of the seller of the house, lots of American dollars were left over. Overlooking the drug smuggler's mansion is yet another mansion, and leading up to it is the best paved road in all of Antigua — even better than the road that was paved for the Queen's visit in 1985 (when the Queen came, all roads that she would travel on were paved anew, so that the Queen might have been left with the impression that riding in a car in Antigua was a pleasant experience). In this mansion lives a woman sophisticated people in Antigua call Evita. She is a notorious woman. She's young and beautiful and the girlfriend of somebody high up in the government. Evita is notorious because her relationship with this high government official has made her the owner of boutiques and property and given her a say in cabinet meetings, and all sorts of other privileges such a relationship would bring a beautiful young woman.

Oh, but by now you are tired of all this looking, and you want to reach your destination — your hotel, your room. You long to refresh yourself; you long to eat some nice lobster, some nice local food. You take a bath, you brush your teeth. You get dressed again; as you get dressed, you look out the window. That water — have you ever seen anything like it? Far out, to the horizon, the colour of the water is navy-blue; nearer, the water is the colour of the North American sky. From there to the shore, the water is pale, silvery, clear, so clear, that you can see its pinkish-white sand bottom. Oh, what beauty! Oh, what beauty! You have never seen anything like this. You are so excited. You breathe shallow. You breathe deep. You see a beautiful boy skimming the water, godlike, on a Windsurfer. You see an incredibly unattractive, fat, pastrylike-fleshed woman enjoying a walk on the beautiful sand, with a man, an incredibly unattractive, fat, pastrylike-fleshed man; you see the pleasure that they're taking in their surroundings. Still standing, looking out the window, you see yourself lying on the beach, enjoying the amazing sun (a sun so powerful and yet so beautiful, the way it is always overhead as if on permanent guard, ready to stamp out any cloud that dares to darken and so empty rain on you and ruin your holiday; a sun that is your personal friend). You see yourself meeting new people (only they are new in a limited way, for they are people just like you). You see yourself eating some delicious, locally grown food. You see yourself, you see yourself .... You must not wonder what exactly happened to the contents of your lavatory when you flushed it. You must not wonder where your bathwater went when you pulled out the stopper. You must not wonder what happened when you brushed your teeth. Oh, it might all end up in the water you are thinking of taking a swim in; the contents of your lavatory might, just might, graze gently against your ankle as you wade carefree in the water, for you see, in Antigua, there is no proper sewage-disposal system. But the Caribbean Sea is very big and the Atlantic Ocean is even bigger; it would amaze even you to know the number of black slaves this ocean has swallowed up. When you sit down to eat your delicious meal, it's better that you don't know that most of what you are eating came off a plane from Miami. And before it got on a plane in Miami, who knows where it came from? A good guess is that it came from a place like Antigua first, where it was grown dirt-cheap, went to Miami, and came back. There is a world of something in this, but I can't go into it right now.

The thing you have always suspected about yourself the minute you become a tourist is true: A tourist is an ugly human being. You are not an ugly person all the time; you are not an ugly person ordinarily; you are not an ugly person day to day. From day to day, you are a nice person. From day to day, all the people who are supposed to love you on the whole do. From day to day, as you walk down a busy street in the large and modern and prosperous city in which you work and love, dismayed, puzzled (a cliché, but only a cliché can explain you) at how alone you feel in this crowd, how awful it is to go unnoticed, how awful it is to go unloved, even as you are surrounded by more people than you could possibly get to know in a lifetime that lasted for millennia, and then out of the corner of your eye you see someone looking at you and absolute pleasure is written all over that person's face, and then you realize that you are not as revolting a presence as you think you are (for that look just told you so). And so, ordinarily, you are a nice person, an attractive person, a person capable of drawing to yourself the affection of other people (people just like you), a person at home in your own skin (sort of; I mean, in a way; I mean, your dismay and puzzlement are natural to you, because people like you just seem to be like that, and so many of the things people like you find admirable about yourselves — the things you think about, the things you think really define you — seem rooted in these feelings): a person at home in your own house (and all its nice house things), with its nice back yard (and its nice back-yard things), at home on your street, your church, in community activities, your job, at home with your family, your relatives, your friends — you are a whole person. But one day, when you are sitting somewhere, alone in that crowd, and that awful feeling of displacedness comes over you, and really, as an ordinary person you are not well equipped to look too far inward and set yourself aright, because being ordinary is already so taxing, and being ordinary takes all you have out of you, and though the words "I must get away" do not actually pass across your lips, you make a leap from being that nice blob just sitting like a boob in your amniotic sac of the modern experience to being a person visiting heaps of death and ruin and feeling alive and inspired at the sight of it; to being a person lying on some faraway beach, your stilled body stinking and glistening in the sand, looking like something first forgotten, then remembered, then not important enough to go back for; to being a person marvelling at the harmony (ordinarily, what you would say is the backwardness) and the union these other people (and they are other people) have with nature. And you look at the things they can do with a piece of ordinary cloth, the things they fashion out of cheap, vulgarly coloured (to you) twine, the way they squat down over a hole they have made in the ground, the hole itself is something to marvel at, and since you are being an ugly person this ugly but joyful thought will swell inside you: their ancestors were not clever in the way yours were and not ruthless in the way yours were, for then would it not be you who would be in harmony with nature and backwards in that charming way? An ugly thing, that is what you are when you become a tourist, an ugly, empty thing, a stupid thing, a piece of rubbish pausing here and there to gaze at this and taste that, and it will never occur to you that the people who inhabit the place in which you have just paused cannot stand you, that behind their closed doors they laugh at your strangeness (you do not look the way they look); the physical sight of you does not please them; you have bad manners (it is their custom to eat their food with their hands; you try eating their way, you look silly; you try eating the way you always eat, you look silly); they do not like the way you speak (you have an accent); they collapse helpless from laughter, mimicking the way they imagine you must look as you carry out some everyday bodily function. They do not like you. *They do not like me?* That thought never actually occurs to you. Still, you feel a little uneasy. Still, you feel a little foolish. Still, you feel a little out of place. But the banality of your own life is very real to you; it drove you to this extreme, spending your days and your nights in the company of people who despise you, people you do not like really, people you would not want to have as your actual neighbour. And so you must devote yourself to puzzling out how much of what you are told is really really true. (Is ground-up bottle glass in peanut sauce really a delicacy around here, or will it do just what you think ground-up bottle glass will do? Is this rare, multicoloured, snout-mouthed fish really an aphrodisiac, or will it cause you to fall asleep permanently?). Oh, the hard work all of this is, and is it any wonder, then, that on your return home you feel the need of a long rest, so that you can recover from your life as a tourist?

That the native does not like the tourist is not hard to explain. For every native of every place is a potential tourist, and every tourist is a native of somewhere. Every native everywhere lives a life of overwhelming and crushing banality and boredom and desperation and depression, and every deed, good and bad, is an attempt to forget this. Every native would like to find a way out, every native would like a rest, every native would like a tour. But some natives — most natives in the world— cannot go anywhere. They are too poor. They are too poor to go anywhere. They are too poor to escape the reality of their lives; and they are too poor to live properly in the place where they live, which is the very place you, the tourist, want to go — so when the natives see you, the tourist, they envy you, they envy your ability to leave your own banality and boredom, they envy your ability to turn their own banality and boredom into a source of pleasure for yourself.

HEROES

URSULA K. Le GUIN

FOR THIRTY YEARS I'VE BEEN FASCINATED by books about the early explorations of the Antarctic, and particularly by the books written by men who were on the expeditions: Scott, Shackleton, Cherry-Garrard, Wilson, Byrd, and so on, all of them not only men of courage and imagination but excellent writers, vivid, energetic, exact, and powerful. As an American I wasn't exposed to the British idolization of Scott that now makes it so chic to sneer at him, and I still feel that I am competent to base my judgment of his character, or Shackleton's, or Byrd's, on their own works and witness, without much reference to the various biases of biographers.

They were certainly heroes to me, all of them. And as I followed them step by frostbitten-toed step across the Ross Ice Barrier and up the Beardmore Glacier to the awful place, the white plateau, and back again, many times, they got into my toes and my bones and my books, and I wrote *The Left Hand of Darkness*, in which a Black man from Earth and an androgynous extraterrestrial pull Scott's sledge through Shackleton's blizzards across a planet called Winter. And fifteen years or so later I wrote a story, "Sur," in which a small group of Latin Americans actually reach the South Pole a year before Amundsen and Scott, but decide not to say anything about it, because if the men knew that they had got there first — they are all women — it wouldn't do. The men would be so let

Ursula K. Le Guin was born in 1929 in California and is the winner of the Hugo, Nebula, and National Book awards. About world-making, she writes, "To make something is to invent it, to discover it, to uncover it, like Michelangelo cutting away the marble that hid the statue. Perhaps we think less often of the proposition reversed, thus: To discover something is to make it. As Julius Caesar said, 'The existence of Britain was uncertain, until I went there.' We can safely assume that the ancient Britons were perfectly certain of the existence of Britain ... {but} as far as Rome, not Britain, is concerned, Caesar invented (invenire, 'to come into, to come upon') Britain. He made it be, for the rest of the world."

down. "We left no footprints, even," says the narrator.

Now, in writing that story, which was one of the pleasantest experiences of my life, I was aware that I was saying some rather hard things about heroism, but I had no desire or intention to debunk or devalue the actual explorers of Antarctica. What I wanted was to join them, fictionally. I had been along with them so many times in their books; why couldn't a few of us, my kind of people, housewives, come along with them in my book ... or even come before them?

These simple little wishes, when they become what people call "ideas" — as in "Where do you get the *ideas* for your stories?" — and when they find themselves in an appropriate nutrient medium such as prose, may begin to grow, to get yeasty, to fizz. Whatever the "idea" of that story was, it has continued to ferment in the dark vats of my mental cellars and is now quite heady, with a marked nose and a complicated aftertaste, like a good '69 Zinfandel.

I wasn't aware of this process until recently, when I was watching the Public Broadcasting series about Shackleton (as well conceived, cast, and produced as the series about Scott and Amundsen was shoddy). There were Ernest Shackleton and his three friends struggling across the abomination of desolation towards the Pole, two days before they had to turn back only ninety-seven miles short of that geometrical *bindu* which they desired so ardently to attain. And the voice-over spoke words from Shackleton's journal: "Man can only do his best. The strongest forces of Nature are arrayed against us." And I sat there and thought, Oh, what nonsense!

That startled me. I had been feeling just as I had always felt for those cold, hungry, tired, brave men, and commiserating them for the bitter disappointment awaiting them — and yet Shackleton's words struck me as disgustingly false, as silly. Why? I had to think it out; and this paper is the process of thinking it out.

"Man can only do his best" — well, all right. They were all men of course, and a long way from the suffragists back home; they honestly believed that "man" includes women, or would have said they did if they had ever thought about it, which I doubt they ever did. I am sure they would have laughed heartily at the proposal that their expedition include women. But still, Man can only do his best; or, to put it in my dialect, people can only do their best; or, as King Yudhisthira says in the great and bitter end of the *Mahabharata*, "By nothing that I do can I attain a goal beyond my reach." That kin whose dog's name is Dharma knows what he is talking about. As did those English explorers, with their fierce sense of duty.

But how about "The strongest forces of Nature are arrayed against us"? Here's the problem. What did you expect, Ernest? Indeed, what did you ask for? Didn't you set it up that way? Didn't you arrange, with vast trouble and expenses that the very strongest "forces of Nature" would be "arrayed against" you and your tiny army?

What is false is the military image; what is foolish is the egoism; what is pernicious is the identification of "Nature" as enemy. We are asked to believe that the Antarctic continent became aware that four Englishmen were penetrating her virgin whiteness and so unleashed upon them the punishing fury of her revenge, the mighty weaponry of wind and blizzard, and so forth and so on. Well, I don't believe it. I don't believe that Nature is either an enemy, or a woman, to humanity. Nobody has ever thought so but Man; and the thought is, to one not Man, no longer acceptable even as a poetic metaphor. Nobody, nothing, "arrayed" any "forces" against Shackleton except Shackleton himself. He created an obstacle to conquer or an enemy to attack; attacked; and was defeated — by what? By himself, having himself created the situation in which his defeat could occur.

Had he reached the Pole he would have said, "I have conquered, I have achieved," in perfectly self-justified triumph. But, forced to retreat, he does not say, "I am defeated"; he blames it on that which is not himself, Nature. If Man wins the battle he starts, he takes the credit for winning, but if he doesn't win, he doesn't lose; "forces arrayed against" him defeat him. Man does not, cannot fail. And Shackleton, speaking for Man, refuses the responsibility for a situation for which he was responsible from beginning to end.

In an even more drastic situation for which he was even more responsible, in his last journal entry Scott wrote:

> We took risks, we knew we took them; things have come out against us, and therefore we have no cause for complaint, but bow to the will of Providence, determined still to do our best to the last.

I have seriously tried to find those words false and silly; I can't do it. Their beauty is no accident.

"Things have come out against us" sounds rather like a projection of fault (like the "forces arrayed against us") but lacks any note of accusation or blame; the underlying image is that of gambling, trusting to luck. "Providence," which is how Scott referred to God, does seem to come in as the "Other," a will opposed to Scott's will as Nature was opposed to Shackleton's; but something you call by the name of Providence is not something you perceive as an opponent or an enemy — indeed, the connotations are maternal: nurturing, sheltering, providing. He takes responsibility for the risks taken, and beyond hope finds duty unalterable: "to do our best to the last." Like Yudhisthira, he knew what "the last" meant. Nothing in me finds this contemptible, and I can't imagine ever finding it contemptible. But I don't know. I have found so many things silly that just a few years ago seemed fine .... Time to bottle the wine: if you leave it too long in the wood it sours and is lost. I don't want to go sour. All I want to do is lose the hero myths so that I can find what is worth admiration.

All right: what I admire in Shackleton, at that moment on the Barrier, is that he turned back. He gave up; he admitted defeat; and he saved his men. Unfortunately he also saved his pride by posturing a bit, playing hero. He couldn't admit that his weakness was his strength; he did the right thing, but said the wrong one. So I go on loving Shackleton, but with the slightest shade of contempt for his having boasted.

But Scott, who did nearly everything wrong, why have I no such contempt for Scott? Why does he remain worthy in my mind of that awful beauty and freedom, my Antarctica? Evidently because he admitted his failure completely — living it through to its end, death. It is as if Scott realized that his life was a story he had to tell, and he had to get the ending right.

This statement may be justly seen as frivolous, trivializing. The death of five people isn't "just a story."

But then, what is a story? And what does one live for? To stay alive, certainly; but only that?

In Amundsen's practical, realistic terms, the deaths of Scott and his four companions were unnecessary, preventable. But then, in what terms was Amundsen's polar journey necessary? It had no justification but nationalism/egoism — "Yah! I'm going to get there first!"

When Scott's party stopped for the last time, the rocks they had collected for the Museum of Natural History were still heavy on the sledge. That is very moving; but I will not use the scientific motives of Scott's expedition to justify his polar journey. It was a mere race too, with no goal but winning. It was when he lost the race that it became a real journey to a real end. And this reality, this value to others, lies in the account he kept.

Amundsen's relation of his polar run is interesting, informative, in some respects admirable. Scott's journal is all that and very much more than that. I would rank it with Woolf's or Pepys's diaries, as a personal record of inestimable value, written by an artist. Scott's temperament was not very well suited to his position as leader; his ambition and intensity drove him to lead, but his inflexibility, vanity, and unpredictability could make his leadership a disaster, for example in his sudden decision to take four men, not three, on the last lap to the Pole, thus oversetting all the meticulous arrangements for supplies. Scott arranged his own defeat, his death, and the death of the four men he was responsible for. He "asked for it." And there were certainly self-destructive elements in his personality. But it would be merely glib to say the he "wanted to fail," and it would miss what I see as the real heroism: what he made of his failure. He took complete responsibility for it. He witnessed truly. He kept on telling the story.

"Unless a grain of wheat fall into the ground and die, it abideth alone; but if it die, it bringeth forth much fruit."

His self-sacrifice was not, I think, deliberate; but his behaviour was sacrificial, rather than heroic. And it was as that unheroic creature, a writer, that he gathered, garnered, saved what could be saved from defeat, suffering, and death. Because he was an artist, his testimony turns mere waste and misery into that useful thing, tragedy.

His companion Edward Wilson, whose paintings are perhaps the finest visual record of Antarctica, kept a diary of the polar journey too. Wilson was a far sweeter, more generous man than Scott, and his diary is very moving, but it has not the power of Scott's — it is not a work of art; it records, but it does not ultimately take responsibility for what happens. Self-absorbed, wilful, obsessed, controlling, Scott was evidently an artist born. He should never have been entrusted with a polar expedition, no doubt. But he was; and he had so fierce a determination to tell his story to the end that he wrote it even as he lay in the tent on the ice dying of cold, starvation, and gangrene among his dead. And so Antarctica is ours. He won it for us.

GROUP MINDS

DORIS LESSING

Doris Lessing was born of British parents in Persia (now Iran) in 1919 and moved to Southern Rhodesia (now Zimbabwe) when she was five. She has lived in England since 1949. She has had more than thirty books published, including the five-volume Children of Violence *series. Her stories reflect what she calls a strong concern for the individual conscience in its relation with the "collective." This is the fourth lecture in her series of talks entitled "Prisons We Choose to Live Inside."*

PEOPLE LIVING IN THE WEST, IN SOCIETIES that we describe as Western, or as the free world, may be educated in many different ways, but they will all emerge with an idea about themselves that goes something like this: I am a citizen of a free society, and that means I am an individual, making individual choices. My mind is my own, my opinions are chosen by me, I am free to do as I will, and at the worst the pressures on me are economic, that is to say I may be too poor to do as I want.

The set of ideas may sound something like a caricature, but it is not so far off how we see ourselves. It is a portrait that may not have been acquired consciously, but is part of a general atmosphere or set of assumptions that influence our ideas about ourselves.

People in the West therefore may go through their entire lives never thinking to analyse this very flattering picture, and as a result are helpless against all kinds of pressures on them to conform in many kinds of ways.

The fact is that we all live our lives in groups — the family, work groups, social, religious and political groups. Very few people indeed are happy as solitaries, and they tend to be seen by their neighbours as peculiar or selfish or worse. Most people cannot stand being alone for long. They are always seeking groups to belong to, and if one group dissolves, they look for another. We are group animals still and there is nothing wrong with that. But as I suggested in the last talk in this series, what is dangerous is not the belonging to a group, or groups, but not understanding the social laws that govern groups and govern us.

When we're in a group, we tend to think as that group does: we may even have joined the group to find "like-minded" people. But we also find our thinking changing because we belong to a group.

It is the hardest thing in the world to maintain an individual dissident opinion, as a member of a group.

It seems to me that this is something we have all experienced — something we take for granted, may never have thought about it. But a great deal of experiment has gone on among psychologists and sociologists on this very theme. If I describe an experiment or two, then anyone listening who may be a sociologist or psychologist will groan, oh God not *again* — for they will have heard of these classic experiments far too often. My guess is that the rest of the people will never have heard of these experiments, never have had these ideas presented to them. If my guess is true, then it aptly illustrates my general thesis, and the general idea (behind these talks) that we (the human race) are now in possession of a great deal of hard information about ourselves, but we do not use it to improve our institutions and therefore our lives.

A typical test, or experiment, on this theme goes like this. A group of people are taken into the researcher's confidence. A minority of one, two, are left in the dark. Some situation demanding measurement or assessment is chosen. For instance, comparing lengths of wood that differ only a little from each other, but enough to be perceptible, or shapes that are almost the same size. The majority in the group — according to instruction — will assert stubbornly that these two shapes or lengths are the same length, or size, while the solitary individual, or the couple, who have not been so instructed will assert that the pieces of wood or whatever are different. But the majority will continue to insist — speaking metaphorically — that black is white, and after a period of exasperation, irritation, even anger, certainly incomprehension, the minority will fall into line. Not always, but nearly always. There are indeed glorious individualists who stubbornly insist on telling the truth as they see it, but most give in to the majority opinion, obey the atmosphere.

When put as baldly, as unflatteringly, as this, reactions tend to be incredulous: "I certainly wouldn't give in, I speak my mind...." But would you?

People who have experienced a lot of groups, who perhaps have observed their own behaviour, may agree that the hardest thing in the world is to stand out against one's group, a group of one's peers. Many agree that among one's most shameful memories are of saying that black is white because other people are saying it.

In other words, we know that this is true of human behaviour, but how do we know it? It is one thing to admit, in a vague uncomfortable sort of way (which probably includes the hope that one will never again be in such a testing situation) but quite another to make that cool step into a kind of objectivity, where one may say: "Right, if that's what human beings are like, myself included, then let's admit it, examine and organize our attitudes accordingly."

This mechanism, of obedience to the group, does not only mean obedience or submission to a small group, or one that is sharply determined, like a religion or political party. It means, too, conforming to those large vague, ill-defined collections of people who may never think of themselves as having a collective mind because they are aware of the differences of opinion — but which, to people from outside, from another culture, seem very minor. The underlying assumptions and assertions that govern the group are never discussed, never challenged, probably never noticed, the main one being precisely this: that it is a group mind, intensely resistant to change, equipped with sacred assumptions about which there can be no discussion.

Since my field is literature, it is there I most easily find my examples. I live in London, and the literary community there would not think of itself as a collective mind, to put it mildly, but that is how I think of it. A few mechanisms are taken for granted enough to be quoted and expected. For instance, what is called "the ten-year rule," which is that usually when a writer dies, her or his work falls out of favour, or from notice, and then comes back again. It is one thing to think vaguely that this is likely to happen, but is it useful? Does it have to happen? Another very noticeable mechanism is the way a writer may fall out of favour for many years — while still alive, be hardly noticed — then suddenly be noticed and praised. An example is Jean Rhys, who lived for many years in the country. She was never mentioned, she might very well have been dead, and most people thought she was. She was in desperate need of friendship and help and did not get it for a long time. Then, due to the efforts of a perspicacious publisher, she finished *The Cruel Sargasso Sea*, and at once as it were became visible again. But — and this is my point — all her previous books, which had been unmentioned and unhonoured, were suddenly remembered and praised. Why were they not praised at all during that long period of neglect? Well, because the collective mind works like that — it is follow-my-leader, people all saying the same thing at the same time.

One can say of course that this is only "the way of the world." But does it have to be? If it does have to be, then at least we could expect it, understand it, and make allowances for it. Perhaps if it is a mechanism that is known to be one then it might be easier for reviewers to be braver and less like sheep in their pronouncements.

Do they have to be so afraid of peer group pressure? Do they really not see how they repeat what each other says?

One may watch how an idea or an opinion, even a phrase, springs up and is repeated in a hundred reviews, criticisms, conversations — and then vanishes. But meanwhile each individual who has bravely repeated this opinion or phrase has been the victim of a compulsion to be like everybody else that has never been analysed — or not by her or him. Though it is easily observable by outsiders.

This is of course the mechanism that journalists rely on when they visit a country. They know if they interview a small sample of a certain kind, or group, or class of people, these two or three citizens will represent all the others, since at any given time, all the people of any group or class or kind will be saying the same things, in the same words.

My experience as Jane Summers illustrates these and many other points. Unfortunately there isn't time here to tell the story properly. I wrote two books under another name, Jane Summers, which were submitted to publishers as if by an unknown author. I did this out of curiosity and to highlight certain aspects of the publishing machine. Also, the mechanisms that govern reviewing. The first, *The Diary of a Good Neighbour*, was turned down by my two main publishers. It was accepted by a third and also by three European publishers. The book was deliberately sent to all the people who regard themselves as experts of my work and they didn't recognize me. Eventually, it was reviewed, as most new novels are, briefly and often patronizingly, and would have vanished forever leaving behind a few fan letters. Because Jane Summers did get fan letters from Britain and the United States, the few people in on the secret were amazed that no one guessed. Then I wrote the second, called *If the Old Could*, and still no one guessed. Now people keep saying to me: "How is it possible that no one guessed? I would have guessed at once." Well, perhaps. And perhaps we're all more dependent on brand names and on packaging than we'd like to think. Just before I came clean, I was asked by an interviewer in the States what I thought would happen. I said that the British literary establishment would be angry and say the books were no good, but that everyone else would be delighted. And this is exactly what happened. I got lots of congratulatory letters from writers and readers who had enjoyed the joke — and very sour and bitchy reviews. However, in France and in Scandinavia the books came out like this: The Diaries of Jane Summers by Doris Lessing. I have seldom had as good reviews as I did in France and in Scandinavia for the Jane Summers books. Of course, one could conclude that the reviewers in France and Scandinavia have no taste but that the British reviewers have.

It has all been very entertaining but it has left me with the feeling, as well, of being sad and embarrassed for my profession. Does everything always have to be so predictable? Do people really have to be such sheep?

Of course, there are original minds, people who do take their own line, who do not fall victim to the need to say, or do, what everyone else does. But they are few. Very few. On them depends the health, the vitality of all our institutions, not only literature, from which I have been drawing my examples.

It has been noticed that there is this ten percent of the population, who can be called natural leaders, who do follow their own minds into decisions and choices. It has been noted to the extent that this fact has been incorporated into instructions for people who run prisons, concentration camps, prisoner of war camps: remove the ten percent, and your prisoners will become spineless and conforming.

Of course, we are back here with the notion of elitism, which is so unfashionable, so unlikeable to the extent that in large areas of politics, even education, the idea that some people may be naturally better equipped than others is resisted. But I will return to the subject of elitism later. Meanwhile, we may note that we all rely on, and we respect, this idea of the lonesome individualist who overturns conformity. It is the recurrent subject of archetypal American films — *Mr. Smith Goes to Washington*, for instance.

Take the way an attitude towards a certain writer or a book will be held by everyone, everyone saying the same things, whether for praise or for blame, until opinion shifts. This shift can be part of some wider social shift. Let us take the Women's Movement, as an example. In London there is a lively, courageous publisher called Virago, run by women. A great many women writers who have been ignored or not taken seriously have been re-evaluated by them. But sometimes the shift is because one person stands out against the prevailing tide of opinion, and the others fall into line behind him, or her, and the new attitude then becomes general.

This mechanism is of course used all the time by publishers. When a new writer, a new novel, has to be launched, the publisher will look for an established writer to praise it. Because one "name" says it is good, the literary editors take notice and the book is launched. It is easy to see this bit of machinery at work in oneself: if someone one respects says such and such a thing is good, when you think it isn't, it is hard to differ. If several people say it is good, then it is correspondently harder.

At a time when one set of attitudes is in the process of changing to another, it is easy to see the hedging-your-bets mechanism. A reviewer will write a piece nicely balanced between one possibility and another. A light, knowing, urbane tone often goes with this. This particular tone is used a great deal on radio and television, when doubtful subjects are under discussion. For example, when it was believed that it was impossible for us to put men on the moon, which is what the Astronomer Royal said a few years before we did. This light, mocking, dismissive tone divorces the speaker from the subject: he or she addresses the listener, the viewer, as if it were over the head of the stupid people who believe that we could put men on the moon, or that there may be monsters in Loch Ness or Lake Champlain, or that ... but fill in your own pet possibility.

Once we have learned to see this mechanism in operation, it can be seen how little of life is free of it. Nearly all the pressures from outside are in terms of group beliefs, group needs, national needs, patriotism and the demands of local loyalties, such as to your city and local groups of all kinds. But more subtle and more demanding — more dangerous — are the pressures from inside, which demand that you should conform, and it is these that are the hardest to watch and to control. If possible.

Many years ago I visited the Soviet Union, during one of their periods of particularly severe literary censorship. The group of writers we met was saying that there was no need for their works to be censored, because they had developed what they called "inner censorship." That they said this with pride shocked us Westerners. What was shocking was that they were so naive about it, cut off as they are from information about psychological and sociological development. This "inner censorship" is what the psychologists call "internalizing" an exterior pressure — such as a parent — and what happens is that a previously resisted and disliked attitude becomes your own.

This happens all the time, and it is often not easy for the victims themselves to know it.

There are other experiments done by psychologists and sociologists that underline that body of experience to which we give the folk-name, "human nature." They are recent; that is to say, done in the last twenty or thirty years. There have been some pioneering and key experiments that have given birth to many others along the same lines — as I said before, over-familiar to the professionals, unfamiliar to most people.

One is known as the Milgram experiment. I have chosen it precisely because it was and is controversial, because it was so much debated, because all the professionals in the field probably groan at the very sound of it. Yet, most ordinary people have never heard of it. The Milgram experiment was prompted by curiosity into how it is that ordinary decent, kindly people, like you and me, will do abominable things when ordered to do them — like the innumerable officials under the Nazis who claimed as an excuse that they were "only obeying orders."

The researcher put into one room people chosen at random who were told that they were taking part in an experiment. A screen divided the room in such a way that they could hear but could not see into the other part. In this second part volunteers sat apparently wired up to a machine that administered electric shocks — with grunts then groans, then screams, then pleas that the experiment should terminate. The person in the first half of the room believed the person in the second half was in fact connected to the machine. He was told that his or her job was to administer increasingly severe shocks according to the instructions of the experimenter and to ignore the cries of pain and pleas from the other side of the screen. Sixty-two percent of the people tested continued to administer shocks up to the 450 volts level. At the 285 volt level the guinea pig had given an agonized scream and become silent. The people administering what they believed were at the best extremely painful doses of electricity were under great stress, but went on doing it. Afterwards most couldn't believe they were capable of such behaviour. Some said: "Well I was only carrying out instructions."

This experiment, like the many others along the same lines, offers us the information that a majority of people, regardless of whether they are black or white, male or female, old or young, rich or poor, will carry out orders, no matter how savage and brutal the orders are. This obedience to authority, in short, is not a property of the Germans, under the Nazis, but a part of general human behaviour. People who have been in a political movement, at times of extreme tension, people who remember how they were at school, will know this anyway... but it is one thing carrying a burden of knowledge around, half-conscious of it, perhaps ashamed of it, hoping it will go away if you don't look too hard, and another saying openly and calmly and sensibly: "Right. This is what we must expect under this and that set of conditions."

Can we imagine this being taught at school, imagine it being taught to children: "If you are in this type or that type of situation, you will find yourself, if you are not careful, behaving like a brute and a savage if you are ordered to do it. Watch out for these situations. You must be on your guard against your own most primitive reactions and instincts."

Another range of experiments is concerned with how children learn best in school. Some results go flat against some of the most cherished assumptions such as, for instance, that they learn best not when "interested" or "stimulated" but when they are bored. But putting that aside — it is known that children learn best from teachers who expect them to learn well. And will do badly if not much is expected from them. Now, we know that in a class of mixed boys and girls, most teachers will — quite unconsciously — spend more time on the boys than the girls, expect much more in scope from the boys, will consistently underestimate the girls. In mixed classes, white teachers will again quite unconsciously — denigrate the non-white children, expect less from them, spend less time on them. These facts, I say, are known — but where are they incorporated, where are they used in schools? In what town is it said to teachers something like this: "As teachers you must become aware of this, that attention is one of your most powerful teaching aids. Attention — the word we give to a certain quality of respect, an alert and heedful interest in a person — is what will feed and nourish your pupils." (To which of course I can already hear the response: "But what would you do if you had thirty children in your class, how much attention could you give to each?") Yes I know, but if these are the facts, if attention is so important, then at some point the people who allot the money for schools and for training programs must, quite simply, put it to themselves like this: children flourish if they are given attention — and their teachers' expectations that they will succeed. Therefore we must pay out enough money to the educators so that enough attention may be provided...

Another range of experiments were carried out extensively in the United States, and for all I know, in Canada too. For instance, a team of doctors cause themselves to be admitted as patients into a mental hospital, unknown to the staff. At once they start exhibiting the symptoms expected of mentally ill people, and start behaving within the range of behaviour described as typical of mentally ill people. The real doctors all, without exception, say they are ill, and classify them in various ways according to the symptoms described by them. It is not the doctors or the nurses who see that these so-called people are quite normal; it is the other patients who see it. They aren't taken in; it is they who can see the truth. It is only with great difficulty that these well people convince the staff that they are well, and obtain their release from the hospital.

Again: a group of ordinary citizens, researchers, cause themselves to be taken into prison, some as if they were ordinary prisoners, a few in the position of warders. Immediately both groups start behaving appropriately: those as warders, with authority, badly treating the prisoners, who for their part, show typical prison behaviour, become paranoid, suspicious, and so forth. Those in the role of warders confessed afterwards they could not prevent themselves enjoying the position of power, enjoying the sensation of controlling the weak. The so-called prisoners could not believe, once they were out, that they had in fact behaved as they had done.

But suppose this kind of thing were taught in schools?

Let us just suppose it, for a moment .... But at once the nub of the problem is laid bare.

Imagine us saying to children: "In the last fifty or so years, the human race has become aware of a great deal of information about its mechanisms; how it behaves, how it must behave under certain circumstances. If this is to be useful, you must learn to contemplate these rules calmly, dispassionately, disinterestedly, without emotion. It is information that will set people free from blind loyalties, obedience to slogans, rhetoric, leaders, group emotions." Well, there it is.

What government, anywhere in the world, will happily envisage its subjects learning to free themselves from governmental and state rhetoric and pressures? Passionate loyalty and subjection to group pressure is what every state relies on. Some, of course, more than others. Khomeini's Iran, and the extreme Islamic sects, the Communist countries, are at one end of the scale. Countries like Norway, whose national day is celebrated by groups of children in fancy dress carrying flowers, singing and dancing, with not a tank or a gun in sight, are at the other. It is interesting to speculate: what country, what nation, when, and where, would have undertaken a program to teach its children to be people to resist rhetoric, to examine the mechanisms that govern them? I can think of only one — America at its birth, in that heady period of the Gettysburg Address. And that time could not have survived the Civil War, for when war starts, countries cannot afford disinterested examination of their behaviour. When a war starts, nations go mad — and have to go mad in order to survive. When I look back at the Second World War, I see something I didn't more than dimly suspect at the time. It was that everyone was crazy. Even people not in the immediate arena of war. I am not talking of the aptitudes for killing, for destruction, which soldiers are taught as part of their training, but a kind of atmosphere, the invisible poison, which spreads everywhere. And then people everywhere begin behaving as they never could in peacetime. Afterwards we look back, amazed. Did I really do that? Believe that? Fall for that bit of propaganda? Think that all our enemies were evil? That all our own nation's acts were good? How could I have tolerated that state of mind, day after day, month after month — perpetually stimulated, perpetually whipped up into emotions that my mind was meanwhile quietly and desperately protesting against?

No, I cannot imagine any nation — or not for long — teaching its citizens to become individuals able to resist group pressures.

And no political party, either. I know a lot of people who are socialists of various kinds, and I try this subject out on them, saying: all governments these days use social psychologists, experts on crowd behaviour, and mob behaviour, to advise them. Elections are stage managed public issues presented according to the rules of mass psychology. The military uses this information. Interrogators, secret services and the police use it. Yet these issues are never even discussed, as far as I am aware, by those parties and groups who claim to represent the people.

On one hand there are governments who manipulate, using expert knowledge and skills, on the other hand people who talk about democracy, freedom, liberty and all the rest of it, as if these values are created and maintained by simply talking about them, by repeating them often enough. How is it that so-called democratic movements don't make a point of instructing their members in the laws of crowd psychology, group psychology?

When I ask this, the response is always an uncomfortable, squeamish reluctance, as if the whole subject is really in very bad taste, unpleasant, irrelevant. As if it will all just go away if it is all just ignored.

So at the moment, if we look around the world, the paradox is that we may see this new information being eagerly studied by governments, the possessors and users of power — studied and put into effect. But the people who say they oppose tyranny literally don't want to know.

The

SCRIBE

PRIMO LEVI

SEVERAL YEARS AGO, IN SEPTEMBER 1984, I bought myself a word processor, that is, a writing tool that returns automatically at the end of a line and makes it possible to insert, cancel, instantaneously change words or entire sentences; in brief, makes it possible to achieve in one leap a finished document, clean, without insertions or corrections. Certainly I'm not the first writer who has decided to take the plunge. Only a year earlier I would have been considered reckless or a snob; today no longer, so fast does electronic time run.

I hasten to add two clarifications. In the first place: whoever wants to or must write can very well continue with his ballpoint or typewriter: my gadget is a luxury, it is amusing, even exciting, but superfluous. Second, to reassure the uncertain and laymen, I

Primo Levi was born in Turin, Italy in 1919. A chemist by trade, he was arrested as a member of the anti-fascist resistance and deported to Auschwitz in 1944. Levi became increasingly convinced that the meaning of the Holocaust was destined to be lost: "The further events fade into the past, the more the construction of convenient truth grows and is perfected." In 1977, he retired from his work to commit himself fully to writing. He died apparently by suicide in 1987.

myself was, indeed still am, as I'm writing here on the screen, a layman. My ideas as to what takes place behind the screen are vague. At first contact, this ignorance of mine humiliated me profoundly; a young man rushed in to reassure me and he has guided me, and to start with he said to me: You belong to the austere generation of humanists who still insist on wanting to understand the world around them. This demand has become absurd: leave everything to habit, and your discomfort will disappear. Consider: do you know or do you think you know how the telephone and television work? And yet you use them every day. And with the exception of a few learned men how many know how their hearts and kidneys work?

Despite this admonition, the first collision with the apparatus was filled with anguish, the anguish of the unknown which for many years I had no longer felt. The computer was delivered to me accompanied by a profusion of manuals; I tried to study them before touching the keys, and I felt lost. It seemed to me that although they were apparently written in Italian, they were in an unknown language; indeed, a mocking and misleading language in which well-known words like "open," "close," and "quit" are used in unusual ways. To be sure, there is a glossary that strives to define them, but proceeds in an opposite direction to that of common dictionaries: these define abstruse terms by having recourse to familiar terms; the glossary would give a new meaning to deceptively familiar terms by having recourse to abstruse terms, and the effect is devastating. How much better it would have been to invent a decisively new terminology for these new things! But once more my young friend intervened and pointed out to me that trying to learn how to use a computer with the help of manuals is as foolish as trying to learn how to swim by reading a treatise without going into the water; indeed, he specified, without even knowing what water is, having heard only a vague talk about it.

So I set about working on two fronts: that is, verifying the instructions of the manuals on the equipment, and immediately the legend of the golem came to mind. It is told that centuries ago a magician-rabbi built a clay automaton with Herculean strength and blind obedience so that it would defend the Jews of Prague from the pogroms; but it remained inert, inanimate, until its maker slipped into its mouth a roll of parchment on which was written a verse from the Torah. At that, the clay golem became a prompt and wise servant: it roamed the streets and kept good guard, but turned to stone again when the parchment was removed. I asked myself whether the builders of my apparatus happened to know this strange story (they certainly are cultivated and even witty people): the computer actually has a mouth, crooked, slightly open in a mechanical grimace. Until I introduce the program floppy disk, the computer doesn't compute anything, it is a lifeless metallic box; but when I turn on the switch a polite luminous signal appears on the small screen: this, in the language of my personal golem, means that he is avid to gulp down the floppy disk. When I have satisfied him, he hums softly, purring like a contented cat, comes alive, and immediately displays his character: he is industrious, helpful, severe with my mistakes, obstinate, and capable of many miracles which I still don't know about and which intrigue me.

Provided he's fed the proper program, he can run a warehouse, or an archive, translate a function in his diagram, compile histograms, even play chess: all undertakings that for the moment do not interest me, indeed, make me melancholy and morose, like the pig who was offered pearls. He can also draw, and this for me is a drawback, of the opposite sort: I hadn't drawn anything since elementary school and now, having available a servomechanism which fabricates for me, custom-made, the images that I cannot draw, and at a command even prints them right in front of my nose, amuses me to an indecent extent and distracts me from more proper uses. I must do violence to myself to "leave" the drawing program and go back to writing.

I have noticed that writing in this way one tends to be prolix. The labour of the past, when stone was carved, led to the "lapidary" style: here the opposite takes place, the manual labour is almost nil, and if one doesn't control oneself one inclines to a wasteful expenditure of words; but there is a providential counter and one must keep one's eye on it.

If I now analyse my initial anxiety, I realize that it was in great part illogical: it contained an old fear of those who write, the fear that the unique, inestimable text worked at so hard, which will give you eternal fame, might be stolen or end up in a manhole. Here you write, the words appear neatly on the screen, well aligned, but they are shadows: they are immaterial, deprived of the reassuring support of the paper. The written speaks out; the screen doesn't; when you're satisfied with the text you "put it on disk," where it becomes invisible. Is it still there, absconding to some little corner of the memory disk, or did you destroy it with some mistaken move? Only after days of experience in *corpore vile* (that is, on false texts, not created but copied), you become convinced that the catastrophe of the lost text was foreseen by the talented gnomes who designed the computer: the destruction of a text requires a manoeuvre which has been made deliberately complicated, and during which the apparatus itself warns you: "Watch out, you're about to commit suicide."

Some twenty-five years ago I wrote a not-very-serious short

story in which after many deontological hesitations, a professional poet decides to buy an electronic Versifier and successfully delegates to it all his activity. My apparatus for the time being does not do as much, but it lends itself in an excellent fashion to the composing of verses, because it permits me to make numerous changes without the page looking dirty or disorderly, and reduces to a minimum the manual effort of writing: "So one observes in me the counterpart." A literary friend of mine objects that in this way one loses the noble joy of the philologist intent on reconstructing, through successive erasures and corrections, the itinerary which leads to the perfection of Leopardi's *Infinite*: he's right, but one can't have everything.

As far as I'm concerned, since I've put bridle, bit, and saddle on my computer, the tedium of being a Dinornis, the survivor of an extinct species, has become attenuated in me: the gloom of being "a survivor of his own time" has almost disappeared. The Greeks said about an uncultured man: "He doesn't know how to read or swim"; today one would have to add, "Nor how to use a computer"; I still don't use it well, I'm not an expert, and I don't know if I ever will be, but I am no longer illiterate. And besides, it is a pleasure to be able to add an item to one's list of memorable "firsts": the first time you saw the sea, passed the border, kissed a woman, gave life to a golem.

The Passing WISDOM of Birds

BARRY LOPEZ

Barry Lopez was born in Port Chester, New York, in 1945 and grew up in southern California and New York City. He is a writer who defends the wild country and its inhabitants. Of his work, Lopez writes, "The continuous work of the imagination to bring what is actual together with what is dreamed is an expression of human evolution. The conscious desire is to achieve a state, even momentarily, that like light is unbounded, nurturing, suffused with wisdom and creation .... Whatever world that is, it lies far ahead. But its outline ... is clear in the landscape, and upon this one can actually hope we will find our way."

I

ON THE EIGHTH OF NOVEMBER, 1519, Hernando Cortés and four hundred Spanish soldiers marched self-consciously out of the city of Iztapalapa, Mexico, and started across the great Iztapalapan Causeway separating the lakes of Xochimilco and Chalco. They had been received the afternoon before in Iztapalapa as demi-gods; but they stared now in disbelief at what lay before them. Reflecting brilliantly on the vast plain of dark water like a landscape of sunlit chalk, its lines sharp as cut stone in the dustless air at 7,200 feet, was the Aztec Byzantium — Tenochtitlán. Mexico City.

It is impossible to know what was in the facile, highly charged mind of Cortés that morning, anticipating his first meeting with the reluctant Montezuma; but Bernal Díaz, who was present, tells us what was on the minds of the soldiers. They asked each other was it real — gleaming Iztapalapa behind them, the smooth causeway beneath their feet, imposing Tenochtitlán ahead? The Spanish had been in the New World for twenty-seven years, but what they discovered in the Valley of Mexico that fall "had never been heard of or seen before, nor even dreamed about" in their world. What astounded them was not, solely, the extent and sophistication of the engineering that divided and encompassed the lakes surrounding Tenochtitlán; nor the evidence that a separate culture, utterly different from their own, pursued a complex life in this huge city. It was the depth and pervasiveness of the natural beauty before their senses.

The day before, they had strolled the spotless streets of Iztapalapa through plots of full-blossomed flowers, arranged in patterns and in colours pleasing to the eye; through irrigated fruit orchards; and into still groves of aromatic trees, like cedar. They sat in the shade of bright cotton awnings in quiet stone patios and marvelled at the well-tended orderliness of the vegetable gardens around them. Roses glowed against the lime-washed walls of the houses like garnets and alexandrites. In the hour before sunset, the cool, fragrant air was filled with the whir and flutter of birds, and lit with birdsong.

That had been Iztapalapa. Mexico City, they thought, even as their leader dismounted that morning with solemn deliberation from that magical creature, the horse, to meet an advancing Montezuma ornately caparisoned in gold and silver and bird feathers — Mexico City, they thought as they approached, could only outdo Iztapalapa. And it did. With Montezuma's tentative welcome they were free to wander in its various precincts. Mexico City confirmed the image of a people gardening with meticulous care and with exquisite attention to line and detail at the edge of nature.

It is clear from Díaz's historical account that the soldiers were stunned by the physical beauty of Tenochtitlán. Venice came to their minds in comparison, because of its canals; but Venice was not as intensely fresh, as well lit as Mexico City. And there was not to be found in Venice, or in Salamanca or Paris for that matter, anything like the great aviaries where thousands of birds — white egrets, energetic wrens and thrushes, fierce accipiters, brilliantly coloured parrots — were housed and tended. They were as captivating, as fabulous, as the displays of flowers: vermilion flycatchers, copper-tailed trogons, green jays, blue-throated hummingbirds, and summer tanagers. Great blue herons, brooding condors.

And throughout the city wild birds nested.

Even Cortés, intensely preoccupied with politics, with guiding a diplomacy of conquest in the region, noticed the birds. He was struck, too, by the affinity of the Mexican people for their gardens and for the measured and intricate flow of water through their city. He took time to write Charles V in Spain, describing it all.

Cortés's men, says Díaz, never seemed to tire of the

arboretums, gardens, and aviaries in the months following their entry into the city. By June 1520, however, Cortés's psychological manipulation of Montezuma and a concomitant arrogance, greed, and disrespect on the part of the Spanish military force had become too much for the Mexicans, and they drove them out. Cortés, relentless and vengeful, returned to the Valley of Mexico eleven months later with a large army and laid siege to the city. Canal by canal, garden by garden, home by home, he destroyed what he had described to Charles V as "the most beautiful city in the world." On June 16, in a move calculated to humiliate and frighten the Mexican people, Cortés set fire to the aviaries.

The grotesqueness and unmitigated violence of Cortés's act has come back to me repeatedly in reading of early European encounters with the landscapes of the New World, like a kind of darkness. The siege of Mexico City was fought barbarously on both sides; and the breathtaking parks and beautiful gardens of Mexico City, of course, stood hard by temples in which human life was regularly offered up to Aztec gods, by priests whose hair was matted with human gore and blood. No human culture has ever existed apart from its dark side. But what Cortés did, even under conditions of war, flies wildly in the face of a desire to find a dignified and honourable relationship with nature. It is an ambitious and vague longing, but one that has been with us for centuries, I think, and which today is a voice heard clearly from many different quarters — political science, anthropology, biology, philosophy. The desire is that, our colonial conquests of the human and natural world finally at an end, we will find our way back to a more equitable set of relationships with all we have subjugated. I say back because the early cultures from which Western civilization evolved, such as the Magdalenian phase of Cro-Magnon culture in Europe, apparently had a less contentious arrangement with nature before the development of agriculture in northern Mesopotamia, and the rise of cities.

The image of Cortés burning the aviaries is not simply for me an image of a kind of destructive madness that lies at the heart of imperialistic conquest; it is also a symbol of a long-term failure of Western civilization to recognize the intrinsic worth of the American landscape, and its potential value to human societies that have since come to be at odds with the natural world. While English, French, and Spanish explorers were cruising the eastern shores of America, dreaming of feudal fiefdoms, gold, and political advantage, the continent itself was, already, occupied in a complex way by more than five hundred different cultures, each of which regarded itself as living in some kind of enlightened intimacy with the land. A chance to rediscover the original wisdom inherent in the myriad sorts of human relationships possible with the nonhuman world, of course, was not of concern to us in the sixteenth century, as it is now, particularly to geographers, philosophers, historians, and ecologists. It would not in fact become clear for centuries that the metaphysics we had thrown out thousands of years before was still intact in tribal America. America offered us the opportunity to deliberate with historical perspective, to see if we wished to reclaim that metaphysics.

The need to reexamine our experience in the New World is, increasingly, a practical need. Contemporary American culture, founded on the original material wealth of the continent, on its timber, ores, and furs, has become a culture that devours the earth. Minerals, fresh water, darkness, tribal peoples, everything the land produces we now consume in prodigious amounts. There are at least two schools of thought on how to rectify this high rate of consumption, which most Western thinkers agree is unsustainable and likely wrongheaded if not disastrous. First, there are technical approaches. No matter how sophisticated or innovative these may be, however, they finally seem only clever or artful adjustments, not solutions. Secondly, we can consider a change in attitude toward nature, adopting a fundamentally different way of thinking about it than we have previously had, perhaps ever had as human beings. The insights of aboriginal peoples are of inestimable value here in rethinking our relationships with the natural world (i.e., in figuring out how to get ourselves back into it); but the solution to our plight, I think, is likely to be something no other culture has ever thought of, something over which !Kung, Inuit, Navajo, Walbiri, and the other traditions we have turned to for wisdom in the twentieth century will marvel at as well.

The question before us is how do we find a viable natural philosophy, one that places us again within the elements of our natural history. The answer, I believe, lies with wild animals.

II

Over the past ten years it has been my privilege to spend time in the field in North America with biologists studying several different kinds of animals, including wolves, polar bears, mountain lions, seals, and whales. Of all that could be said about this exercise, about people watching animals, I would like to restrict myself to but one or two things. First, although such studies are scientific they are conducted by human beings whose individual speculations may take them out beyond the bounds of scientific inquiry. The animals they scrutinize may draw them back into an older, more intimate and less rational association with the local landscape. In this frame of mind, they may privately begin to question the methodology of Western science, especially its purported objectivity and its troublesome lack of heart. It may seem to them incapable of addressing questions they intuit are crucial. Even as they perceive its flaws, however, scientists continue to offer such studies as a dependable source of reliable information — and they are. Science's flaws as a tool of inquiry are relatively minor, and it is further saved by its strengths.

Science's strength lies with its rigour and objectivity, and it is undoubtedly as rigorous as any system available to us. Even with its flaws (its failure, for example, to address disorderly or idiosyncratic behaviour) field biology is as strong and reliable in its way as the collective wisdom of a hunting people actively involved with the land. The highest order of field work being done in biology today, then, from an elucidation of the way polar bears hunt ringed seals to working out the ecology of night-flying moths pollinating agaves in the Mojave Desert, forms part of the foundation for a modern realignment with the natural world. (The other parts of the foundation would include work done by anthropologists among hunter-gatherer people and studies by natural geographers; philosophical work in the tradition of Aldo Leopold and Rachel Carson; and the nearly indispensable element of personal experience.)

I often search out scientific reports to read; many are based on years of research and have been patiently thought through. Despite my regard, however, I read with caution, for I cannot rid myself of the thought that, even though it is the best theoretical approach we have, the process is not perfect. I have participated in some of this type of work and know that innocent mistakes are occasionally made in the data. I understand how influential a misleading coincidence can become in the overall collection of data; how unconsciously the human mind can follow a teasing parallel. I am cautious, too, for even more fundamental reasons. It is hard to say exactly what any animal is doing. It is impossible to know when or where an event in an animal's life begins or ends. And our human senses confine us to realms that may contain only a small part of the information produced in an event. Something critical could be missing and we would not know. And as far as the experiments are concerned, although we can design and carry out thousands of them, no animal can ever be described as the sum of these experiments. And, finally, though it is possible to write precisely about something, this does not automatically mean one is accurate.

The scientific approach is flawed, therefore, by its imposition of a subjective framework around animal behaviour; but it only fails, really, because it is incomplete. We would be rash, using this approach exclusively, to claim to understand any one animal, let alone the environment in which that animal is evolving. Two remedies to this dilemma of the partially perceived animal suggest themselves. One, obviously, is to turn to the long-term field observations of non-Western cultural traditions. These non-Aristotelian, non-Cartesian, non-Baconian views of wild animals are stimulating, challenging, and, like a good bibliography, heuristic, pointing one toward discovery. (They are also problematic in that, for example, they do not take sufficiently into account the full range of behaviour of migratory animals and they have a highly nonlinear [though ultimately, possibly, more correct] understanding of population biology.)

A second, much less practised remedy is to cultivate within ourselves a sense of mystery — to see that the possibilities for an expression of life in any environment, or in any single animal, are larger than we can predict or understand, and that this is all right. Biology should borrow here from quantum physics, which accepts the premise that under certain circumstances, the observer can be deceived. Quantum physics, with its ambiguous particles and ten-dimensional universes, is a branch of science that has in fact returned us to a state of awe with nature, without threatening our intellectual capacity to analyse complex events.

If it is true that modern people desire a new relationship with the natural world, one that is not condescending, manipulative, and purely utilitarian; and if the foundation upon which the relationship is to be built is as I suggest — a natural history growing largely out of science and the insights of native peoples — then a staggering task lies before us.

The initial steps to be taken seem obvious. First, we must identify and protect those regions where landscapes largely undisturbed by human technology remain intact. Within these ecosystems lie blueprints for the different patterns of life that have matured outside the pervasive influence of myriad Western technologies (though no place on earth has escaped their influence entirely). We can contemplate and study endlessly the natural associations here, and draw from these smaller universes a sophisticated wisdom about process and event, and about evolution. Second, we need to subscribe a great public support to the discipline of field biology. Third, we need to seek an introduction to the reservoirs of intelligence that native cultures have preserved in both oral tradition and in their personal experience with the land, the highly complex detail of a way of life not yet torn entirely from the fabric of nature.

We must, too, look out after the repositories of our own long-term cultural wisdom more keenly. Our libraries, which preserve the best of what we have to say about ourselves and nature, are under siege in an age of cost-benefit analysis. We need to immerse ourselves thoughtfully, too, in what is being written and produced on tape and film, so that we become able to distinguish again between truthful expression and mere entertainment. We need to do this not only for our own sake but so that our children, who too often have only the half-eclipsed lives of zoo animals or the contrived dramas of television wildlife adventure before them, will know that this heritage is disappearing and what we are apt to lose with it.

What disappears with a debasement of wild landscapes is more than genetic diversity, more than a homeland for Henry Beston's "other nations," more, to be perfectly selfish, than a source of future medical cures for human illness or a chance for personal revitalization on a wilderness trip. We stand to lose the focus of our ideals. We stand to lose our sense of dignity, of compassion, even our sense of what we call God. The philosophy of nature we set aside eight thousand years ago in the Fertile Crescent we can, I think, locate again and greatly refine North America. The New World is a landscape still overwhelming in the vigour of its animals and plants, resonant with mystery. It encourages, still, an enlightened response toward indigenous cultures that differ from our own, whether Aztecan, Lakotan, lupine, avian, or invertebrate. By broadening our sense of the intrinsic worth of life and by cultivating respect for other ways of moving toward perfection, we may find a sense of resolution we have been looking for, I think, for centuries.

Two practical steps occur to me. Each by itself is so small I hesitate to set it forth; but to say nothing would be cowardly, and both appear to me to be reasonable, a beginning. They also acknowledge an obvious impediment: to bridge the chasm between a colonial attitude toward the land and a more filial relationship with it takes time. The task has always been, and must be, carried into the next generation.

    The first thought I would offer is that each university and college establish the position of university naturalist, a position to be held by a student in his or her senior year and passed on at the end of the year to another student. The university naturalist would be responsible for establishing and maintaining a natural history of the campus, would confer with architects and grounds keepers, escort guests, and otherwise look out after the nonhuman elements of the campus, their relationships to human beings, and the preservation of this knowledge. Though the position itself might be honourable and unsalaried, the student would receive substantial credit for his or her work and would be provided with a budget to conduct research, maintain a library, and produce an occasional paper. Depending on his or her gifts and personality, the university naturalist might elect to teach a course or to speak at some point during the academic year. In concert with the university archivist and university historian, the university naturalist would seek to protect the relationships-in-time that define a culture's growth and ideals.

    A second suggestion is more difficult to implement, but no less important than a system of university naturalists. In recent years several American and British publishers have developed plans to reprint in an extended series classic works of natural history. These plans should be pursued; the list of books should include not only works of contemporary natural history but early works by such people as Thomas Nuttal and William Bartram, so that the project has historical depth. It should also include books by nonscientists who have immersed themselves "beyond reason" in the world occupied by animals and who have emerged with stunning reports, such as J.A. Baker's *The Peregrine*. And books that offer us a resounding and affecting vision of the landscape, such as John Van Dyke's *The Desert*. It should also include the writing of anthropologists who have worked, or are working, with the native peoples of North America to define an indigenous natural history, such as Richard Nelson's *Make Prayers to the Raven*. And a special effort should be made to unearth those voices that once spoke eloquently for parts of the country the natural history of which is now too often overlooked, or overshadowed, by a focus on western or northern North American ecosystems: the pine barrens of New Jersey, the Connecticut River Valley, the White Mountains of New Hampshire, the remnant hardwood forests of Indiana and Ohio, the Outer Banks, the relictual prairies of Texas, and the mangrove swamps and the piney woods of Georgia.

Such a collection, it seems to me, should be assembled with several thoughts in mind. It should be inexpensive so that the books can fall easily into the hands of young people. It should document the extraordinary variety of natural ecosystems in North America, and reflect the great range of dignified and legitimate human response to them. And it should make clear that human beings belong in these landscapes, that they, too, are a part of the earth's natural history.

III

The image I carry of Cortés setting fire to the aviaries in Mexico City that June day in 1521 is an image I cannot rid myself of. It stands, in my mind, for a fundamental lapse of wisdom in the European conquest of America, an underlying trouble in which political conquest, personal greed, revenge, and national pride outweigh what is innocent, beautiful, serene, and defenceless — the birds. The incineration of these creatures 450 years ago is not something that can be rectified today. Indeed, one could argue, the same oblivious irreverence is still with us, among those who would ravage and poison the earth to sustain the economic growth of Western societies. But Cortés's act can be transcended. It is possible to fix in the mind that heedless violence, the hysterical cry of the birds, the stench of death, to look it square in the face and say that there is more to us than this, this will not forever distinguish us from among the other cultures. It is possible to imagine that on the far side of the Renaissance and the Enlightenment we can recover the threads of an earlier wisdom.

Again I think of the animals, because of the myriad ways in which they have helped us since we first regarded each other differently. They offered us early models of rectitude and determination in adversity, which we put in stories. The grace of a moving animal, in some ineluctable way, kindles still in us a sense of imitation. They continue to produce for us a sense of the Other: to encounter a truly wild animal on its own ground is to know the defeat of thought, to feel reason overpowered. The animals have fed us; and the cultures of the great hunters particularly — the bears, the dogs, and the cats — have provided the central metaphors by which we have taken satisfaction in our ways and explained ourselves to strangers.

Cortés's soldiers, on their walks through the gleaming gardens of Tenochtitlán, would have been as struck by the flight paths of songbirds as we are today. In neither a horizontal nor a vertical dimension do these pathways match the line and scale of human creation, within which the birds dwell. The corridors they travel are curved around some other universe. When the birds passed over them, moving across the grain of sunlight, the soldiers must have stopped occasionally to watch, as we do. It is the birds' independence from predictable patterns of human design that draws us to them. In the birds' separate but related universe we are able to sense hope for ourselves. Against a background of the familiar, we recognize with astonishment a new pattern.

In such a moment, pausing to take in the flight of a flock of birds passing through sunshine and banking gracefully into a grove of trees, it is possible to move beyond a moment in the Valley of Mexico, when we behaved as though we were insane.

ZAABALAWI

FINALLY I BECAME CONVINCED THAT I HAD TO find the Sheikh Zaabalawi.

The first time I heard his name had been in a song:

> Oh what's become of the world, Zaabalawi?
> They've turned it upside down and taken away its taste.

It had been a popular song in my childhood, and one day it had occurred to me to demand of my father, in the way children have of asking endless questions:

"Who is Zaabalawi?"

He had looked at me hesitantly as

Naguib Mahfouz was born in Cairo in 1911 and won the Nobel Prize for Literature in 1988. He reflects: "Why did I begin to write? For pleasure ... to satisfy a deep dark force ... nothing coming from the outside pushed me toward this act. Later, new reasons completed this desire to write: a desire to carry weight ... to receive retribution through one's work ... to show the other the principles wrought through a search ... simply so that people read me. Today I can hardly distinguish between living and writing."

though doubting my ability to understand the answer. However, he had replied, "May his blessing descend upon you, he's a true saint of God, a remover of worries and troubles. Were it not for him I would have died miserably —"

In the years that followed, I heard my father many a time sing the praises of this good saint and speak of the miracles he performed. The days passed and brought with them many illnesses, for each one of which I was able, without too much trouble and at a cost I could afford, to find a cure, until I became afflicted with that illness for which no one possesses a remedy. When I tried everything in vain and was overcome by despair, I remembered by chance what I had heard in my childhood: Why, I asked myself, should I not seek out Sheikh Zaabalawi? I recollected my father saying that he had made his acquaintance in Khan Gaafar at the house of Sheikh Qamar, one of those sheikhs who practised law in the religious courts, and so I took myself off to his house. Wishing to make sure that he was still living there, I made inquiries of a vendor of beans whom I found in the lower part of the house.

"Sheikh Qamar!" he said, looking at me in amazement. "He left the quarter ages ago. They say he's now living in Garden City and has his office in al-Azhar Square."

I looked up the office address in the telephone book and immediately set off to the Chamber of Commerce Building, where it was located. On asking to see Sheikh Qamar, I was ushered into a room just as a beautiful woman with a most intoxicating perfume was leaving it. The man received me with a smile and motioned me toward a fine leather-upholstered chair. Despite the thick soles of my shoes, my feet were conscious of the lushness of the costly carpet. The man wore a lounge suit and was smoking a cigar; his manner of sitting was that of someone well satisfied both with himself and with his worldly possessions. The look of warm welcome he gave me left no doubt in my mind that he thought me a prospective client, and I felt acutely embarrassed at encroaching upon his valuable time.

"Welcome!" he said, prompting me to speak.

"I am the son of your old friend Sheikh Ali al-Tatawi," I answered so as to put an end to my equivocal position.

A certain langour was apparent in the glance he cast at me; the langour was not total in that he had not as yet lost all hope in me.

"God rest his soul," he said. "He was a fine man."

The very pain that had driven me to go there now prevailed upon me to stay.

"He told me," I continued, "of a devout saint named Zaabalawi whom he met at Your Honour's. I am in need of him, sir, if he be still in the land of the living."

The langour became firmly entrenched in his eyes, and it would have come as no surprise if he had shown the door to both me and my father's memory.

"That," he said in the tone of one who has made up his mind to terminate the conversation, "was a very long time ago and I scarcely recall him now."

Rising to my feet so as to put his mind at rest regarding my intention of going, I asked, "Was he really a saint?"

"We used to regard him as a man of many miracles."

"And where could I find him today?" I asked, making another move toward the door.

"To the best of my knowledge he was living in the Birwagi Residence in al-Azhar," and he applied himself to some papers on his desk with a resolute movement that indicated he would not open his mouth again. I bowed my head in thanks, apologized several times for disturbing him, and left the office, my head so buzzing with embarrassment that I was oblivious to all sounds around me.

I went to Birwagi Residence, which was situated in a thickly populated quarter. I found that time had so eaten away at the building that nothing was left of it save an antiquated facade and a courtyard that, despite being supposedly in the charge of a caretaker, was being used as a rubbish dump. A small, insignificant fellow, a mere prologue to a man, was using the covered entrance as a place for the sale of old books on theology and mysticism.

When I asked him about Zaabalawi, he peered at me through narrow, inflamed eyes and said in amazement, "Zaabalawi! Good heavens, what a time ago that was! Certainly he used to live in this house when it was habitable. Many were the times he would sit with me talking of bygone days, and I would be blessed by his holy presence. Where, though, is Zaabalawi today?"

He shrugged his shoulders sorrowfully and soon left me, to attend to an approaching customer. I proceeded to make inquiries of many shopkeepers in the district. While I found that a large number of them had never even heard of Zaabalawi, some, though recalling nostalgically the pleasant times they had spent with him, were ignorant of his present whereabouts, while others openly made fun of him, labelled him a charlatan, and advised me to put myself in the hands of a doctor — as though I had not already done so. I therefore had no alternative but to return disconsolately home.

With the passing of days like motes in the air, my pains grew so severe that I was sure I would not be able to hold out much longer. Once again I fell to wondering about Zaabalawi and clutching at the hope his venerable name stirred within me. Then it occurred to me to seek the help of the local sheikh of the district; in fact, I was surprised I had not thought of this to begin with. His office was in the nature of a small shop, except that it contained a desk and a telephone, and I found him sitting at his desk, wearing a jacket over his striped galabeya. As he did not interrupt his conversation with a man sitting beside him, I stood waiting till the man had gone. The sheikh then looked up at me coldly. I told myself that I should win him over by the usual methods, and it was not long before I had him cheerfully inviting me to sit down.

"I'm in need of Sheikh Zaabalawi," I answered his inquiry as to the purpose of my visit.

He gazed at me with the same astonishment as that shown by those I had previously encountered.

"At least," he said, giving me a smile that revealed his gold teeth, "he is still alive. The devil of it is, though, he has no fixed abode. You might well bump into him as you go out of here, on the other hand you might spend days and months in fruitless searching."

"Even you can't find him!"

"Even I! He's a baffling man, but I thank the Lord that he's still alive!"

He gazed at me intently, and murmured, "It seems your condition is serious."

"Very."

"May God come to your aid! But why don't you go about it systematically?" He spread out a sheet of paper on the desk and drew on it with unexpected speed and skill until he had made a full plan of the district, showing all the various quarters, lanes, alleyways, and squares. He looked at it admiringly and said, "These are dwelling-houses, here is the Quarter of the Perfumers, here the Quarter of the Coppersmiths, the Mouski, the police and fire stations. The drawing is your best guide. Look carefully in the cafés, the places where the dervishes perform their rites, the mosques and prayer-rooms, and the Green-Gate, for he may well be hidden among the beggars and be indistinguishable from them. Actually, I myself haven't seen him for years, having been somewhat preoccupied with the cares of the world, and was only brought back by your inquiry to those most exquisite times of my youth."

I gazed at the map in bewilderment. The telephone rang, and he took up the receiver.

"Take it," he told me, generously, "We're at your service."

Folding up the map, I left and wandered off through the

quarter, from square to street to alleyway, making inquiries of everyone I felt was familiar with the place. At last the owner of a small establishment for ironing clothes told me, "Go to the calligrapher Hassanein in Umm al-Ghulam — they were friends."

I went to Umm al-Ghulam, where I found old Hassanein working in a deep, narrow shop full of signboards and jars of colour. A strange smell, a mixture of glue and perfume, permeated its every corner. Old Hassanein was squatting on a sheepskin rug in front of a board propped against the wall; in the middle of it he had inscribed the word "Allah" in silver lettering. He was engrossed in embellishing the letters with prodigious care. I stood behind him, fearful of disturbing him or breaking the inspiration that flowed to his masterly hand. When my concern at not interrupting him had lasted some time, he suddenly inquired with unaffected gentleness, "Yes?"

Realizing that he was aware of my presence, I introduced myself. "I've been told that Sheikh Zaabalawi is your friend; I'm looking for him," I said.

His hand came to a stop. He scrutinized me in astonishment. "Zaabalawi! God be praised!" he said with a sigh.

"He is a friend of yours, isn't he?" I asked eagerly.

"He was, once upon a time. A real man of mystery: he'd visit you so often that people would imagine he was your nearest and dearest, then would disappear as though he'd never existed. Yet saints are not to be blamed."

The spark of hope went out with the suddenness of a lamp snuffed by a power-cut.

"He was so constantly with me," said the man, "that I felt him to be a part of everything I drew. But where is he today?"

"Perhaps he is still alive?"

"He's alive without a doubt ... he had impeccable taste, and it was due to him that I made my most beautiful drawings."

"God knows," I said, in a voice almost stifled by the dead ashes of hope, "how dire my need for him is, and no one knows better than you of the ailments in respect to which he is sought."

"Yes, yes. May God restore you to health. He is in truth, as is said of him, a man, and more ..."

Smiling broadly, he added, "And his face possesses an

unforgettable beauty. But where is he?"

Reluctantly I rose to my feet, shook hands, and left. I continued wandering eastward and westward through the quarter, inquiring about Zaabalawi from everyone who, by reason of age or experience, I felt might be likely to help me. Eventually I was informed by a vendor of lupine that he had met him a short while ago at the house of Sheikh Gad, the well-known composer. I went to the musician's house in Tabakshiyya, where I found him in a room tastefully furnished in the old style, its walls redolent with history. He was seated on a divan, his famous lute beside him, concealing within itself the most beautiful melodies of our age, while somewhere from within the house came the sound of pestle and mortar and the clamour of children. I immediately greeted him and introduced myself, and was put at my ease by the unaffected way in which he received me. He did not ask, either in words or gesture, what had brought me, and I did not feel that he even harboured any such curiosity. Amazed at his understanding and kindness, which boded well, I said, "O Sheikh Gad, I am an admirer of yours, having long been enchanted by the rendering of your songs."

"Thank you," he said with a smile.

"Please excuse my disturbing you," I continued timidly, "but I was told that Zaabalawi was your friend, and I am in urgent need of him."

"Zaabalawi!" he said, frowning in concentration. "You need him? God be with you, for who knows, O Zaabalawi, where you are."

"Doesn't he visit you?" I asked eagerly.

"He visited me some time ago. He might well come right now; on the other hand I mighn't see him till death!"

I gave an audible sigh and asked, "What made him like that?"

The musician took up his lute. "Such are saints or they would not be saints," he said laughing.

"Do those who need him suffer as I do?"

"Such suffering is part of the cure!"

He took up the plectrum and began plucking soft strains from the strings. Lost in thought, I followed his movements. Then, as though addressing myself, I said, "So my visit has been in vain."

He smiled, laying his cheek against the side of the lute. "God forgive you," he said, "for saying such a thing of a visit that has caused me to know you and you me!"

I was much embarrassed and said apologetically, "Please forgive me; my feelings of defeat made me forget my manners."

"Do not give in to defeat. This extraordinary man brings fatigue to all who seek him. It was easy enough with him in the old days, when his place of abode was known. Today, though, the world has changed, and after having enjoyed a position attained only by potentates, he is now pursued by the police on a charge of false pretences. It is therefore no longer an easy matter to reach him, but have patience and be sure that you will do so."

He raised his head from the lute and skilfully fingered the opening bars of a melody. Then he sang:

> "I make lavish mention, even though I blame myself,
> of those I love,
> For the stories of the beloved are my wine."

With a heart that was weary and listless, I followed the beauty of the melody and the singing.

"I composed the music to this poem in a single night," he told me when he had finished. "I remember that it was the eve of the Lesser Bairam. Zaabalawi was my guest for the whole of that night, and the poem was of his choosing. He would sit for a while just where you are, then would get up and play with my children as though he were one of them. Whenever I was overcome by weariness or inspiration failed me, he would punch me playfully in the chest and joke with me, and I would bubble over with melodies, and thus I continued working till I finished the most beautiful piece I have ever composed."

"Does he know anything about music?"

"He is the epitome of things musical. He has an extremely beautiful speaking voice, and you have only to hear him to want to burst into song and to be inspired to creativity...."

"How was it that he cured those diseases before which men are powerless?"

"That is his secret. Maybe you will learn it when you meet."

But when would that meeting occur? We relapsed into

silence, and the hubbub of children once more filled the room.

Again the sheikh began to sing. He went on repeating the words "and I have a memory of her" in different and beautiful variations until the very walls danced in ecstasy. I expressed my wholehearted admiration, and he gave me a smile of thanks. I then got up and asked permission to leave, and he accompanied me to the front door. As I shook him by the hand, he said, "I hear that nowadays he frequents the house of Hagg Wanas al-Damanhouri. Do you know him?"

I shook my head, though a modicum of renewed hope crept into my heart.

"He is a man of private means," the sheikh told me, "who from time to time visits Cairo, putting up at some hotel or other. Every evening, though, he spends at the Negma Bar in Alfi Street."

I waited for nightfall and went to the Negma Bar. I asked a waiter about Hagg Wanas, and he pointed to a corner that was semisecluded because of its position behind a large pillar with mirrors on all four sides. There I saw a man seated alone at a table with two bottles in front of him, one empty, the other two-thirds empty. There were no snacks or food to be seen, and I was sure that I was in the presence of a hardened drinker. He was wearing a loosely flowing silk galabeya and a carefully wound turban; his legs were stretched out toward the base of the pillar, and as he gazed into the mirror in rapt contentment, the sides of his face, rounded and handsome despite the fact that he was approaching old age, were flushed with wine. I approached quietly till I stood but a few feet away from him. He did not turn toward me or give me any indication that he was aware of my presence.

"Good evening, Mr. Wanas," I greeted him cordially.

He turned toward me abruptly, as though my voice had roused him from slumber, and glared at me in disapproval. I was about to explain what had brought me when he interrupted in an almost imperative tone of voice that was nonetheless not devoid of an extraordinary gentleness, "First, please sit down, and second, please get drunk!"

I opened my mouth to make my excuses, but, stopping up his ears with his fingers, he said, "Not a word till you do what I say."

I realized I was in the presence of a capricious drunkard and told myself that I should at least humour him a bit. "Would you permit me to ask one question?" I said with a smile, sitting down.

Without removing his hands from his ears he indicated the bottle. "When engaged in a drinking bout like this, I do not allow any conversation between myself and another unless, like me, he is drunk, otherwise all propriety is lost and mutual comprehension is rendered impossible."

I made a sign indicating that I did not drink.

"That's your lookout," he said offhandedly. "And that's my condition!"

He filled me a glass, which I meekly took and drank. No sooner had the wine settled in my stomach than it seemed to ignite. I waited patiently till I had grown used to its ferocity, and said, "It's very strong, and I think the time has come for me to ask you about —"

Once again, however, he put his fingers in his ears. "I shan't listen to you until you're drunk!"

He filled up my glass for a second time. I glanced at it in trepidation; then, overcoming my inherent objection, I drank it down at a gulp. No sooner had the wine come to rest inside me than I lost all willpower. With the third glass, I lost my memory, and with the fourth the future vanished. The world turned round about me, and I forgot why I had gone there. The man leaned toward me attentively, but I saw him — saw everything — as a mere meaningless series of coloured planes. I don't know how long it was before my head sank down into the arm of the chair and I plunged into a deep sleep. During it, I had a beautiful dream the like of which I had never experienced. I dreamed that I was in an immense garden surrounded on all sides by luxuriant trees, and the sky was nothing but stars seen between the entwined branches, all enfolded in an atmosphere like that of sunset or a sky overcast with cloud. I was lying on a small hummock of jasmine petals, more of which fell upon me like rain, while the lucent spray of a fountain unceasingly sprinkled the crown of my head and my temples. I was in a state of deep contentedness, of ecstatic serenity. An orchestra of warbling and cooing played in my ear. There was an extraordinary sense of harmony between me and my inner self, and between the two of us and the world, everything being in its rightful place, without discord or distortion. In the whole world there was no single reason for speech or movement, for the universe moved in a rapture of ecstasy. This lasted but a short while. When I opened my eyes, consciousness struck at me like a policeman's fist, and I saw Wanas al-Damanhouri peering at me with concern. Only a few drowsy customers were left in the bar.

"You have slept deeply," said my companion. "You were obviously hungry for sleep."

I rested my heavy head in the palms of my hands. When I took them away in astonishment and looked down at them, I found that they glistened with drops of water.

"My head's wet," I protested.

"Yes, my friend tried to rouse you," he answered quietly.

"Somebody saw me in this state?"

"Don't worry, he is a good man. Have you not heard of Sheikh Zaabalawi?"

"Zaabalawi!" I exclaimed, jumping to my feet.

"Yes," he answered in surprise. "What's wrong?"

"Where is he?"

"I don't know where he is now. He was here and then he left."

I was about to run off in pursuit but I found I was more exhausted than I had imagined. Collapsed over the table, I cried out in despair, "My sole reason for coming to you was to meet him! Help me catch up to him or send someone after him."

The man called a vendor of prawns and asked him to seek out the sheikh and bring him back. Then he turned to me. "I didn't realize you were afflicted. I'm very sorry...."

"You wouldn't let me speak," I said irritably.

"What a pity! He was sitting on this chair next to you the whole time. He was playing with a string of jasmine petals he has around his neck, a gift from one of his admirers, then, taking pity on you, he began to sprinkle some water on your head to bring you around."

"Does he meet you here every night?" I asked, my eyes not leaving the doorway through which the vendor of prawns had left.

"He was with me tonight, last night, and the night before that, but before that I hadn't seen him for a month."

"Perhaps he will come tomorrow," I answered with a sigh.

"Perhaps."

"I am willing to give any money he wants."

Wanas answered sympathetically, "The strange thing is that he is not open to such temptations, yet he will cure you if you meet him."

"Without charge?"

"Merely on sensing that you love him."

The vendor of prawns returned, having failed in his mission.

I recovered some of my energy and left the bar, albeit unsteadily. At every street corner I called out "Zaabalawi!" in the vague hope that I would be rewarded with an answering shout. The street boys turned contemptuous eyes on me till I sought refuge in the first available taxi.

The following evening I stayed up with al-Damanhouri till dawn, but the sheikh did not put in an appearance. Wanas informed me that he would be going away to the country and would not be returning to Cairo until he had sold the cotton crop.

I must wait, I told myself; I must train myself to be patient. Let me content myself with having made certain of the existence of Zaabalawi, and even of his affection for me, which encourages me to think that he will be prepared to cure me if a meeting takes place between us.

Sometimes, however, the long delay wearied me. I would become beset by despair and would try to persuade myself to dismiss him from my mind completely. How many weary people in this life know him not or regard him as a mere myth! Why, then, should I torture myself about him in this way?

No sooner, however, did my pains force themselves upon me than I would begin to think about him, asking myself when I would be fortunate enough to meet him. The fact that I ceased to have any news of Wanas and was told he had gone to live abroad did not deflect me from my purpose; the truth of the matter was that I had become fully convinced that I had to find Zaabalawi.

Yes, I have to find Zaabalawi.

Monologue of
ISABEL
watching it rain in Macondo

GABRIEL GARCIA MARQUEZ

Gabriel Garcia Marquez was born in 1922 in Aracataca, Colombia and won the Nobel Prize for Literature in 1982.
He has attributed his style to the way his grandmother told her stories:
"She told things that sounded supernatural and fantastic, but she told them with complete naturalness. What was most important was the expression she had on her face. She did not change her expression at all when telling her stories ...
I discovered that what I had to do was believe in my stories myself and write them with the same expression with which my grandmother told them: with a brick face."

WINTER FELL ONE SUNDAY WHEN people were coming out of church. Saturday night had been suffocating. But even on Sunday morning nobody thought it would rain. After mass, before we women had time to find the catches on our parasols, a thick, dark wind blew, which with one broad, round swirl swept away the dust and hard tinder of May. Someone next to me said: "It's a water wind." And I knew it even before then. From the moment we came out onto the church steps I felt shaken by a slimy feeling in my stomach. The men ran to the nearby houses with one hand on their hats and a handkerchief in the other, protecting themselves against the wind and the dust storm. Then it rained. And the sky was a grey, jellyish substance that flapped its wings a hand away from our heads.

During the rest of the morning my stepmother and I were sitting by the railing, happy that the rain would revive the thirsty rosemary and nard in the flowerpots after seven months of intense summer and scorching dust. At noon the reverberation of the earth stopped and a smell of turned earth, of awakened and renovated vegetation mingled with the cool and healthful odour of the rain in the rosemary. My father said at lunchtime: "When it rains in May, it's a sign that there'll be good tides." Smiling, crossed by the luminous thread of the new season, my stepmother told me: "That's what I heard in the sermon." And my father smiled. And he ate with a good appetite and even let his food digest leisurely beside the railing, silent, his eyes closed, but not sleeping, as if to think that he was dreaming while awake.

It rained all afternoon in a single tone. In the uniform and peaceful intensity you could hear the water fall, the way it is when you travel all afternoon on a train. But without our noticing it, the rain was penetrating too deeply into our senses. Early Monday morning, when we closed the door to avoid the cutting, icy draft that blew in from the courtyard, our senses had been filled with rain. And on Monday morning they had overflowed. My stepmother and I went back to look at the garden. The harsh grey earth of May had been changed overnight into a dark, sticky substance like cheap soap. A trickle of water began to run off the flowerpots. "I think they had more than enough water during the night," my stepmother said. And I noticed that she had stopped smiling and that her joy of the previous day had changed during the night into a lax and tedious seriousness. "I think you're right," I said. "It would be better to have the Indians put them on the veranda until it stops raining." And that was what they did, while the rain grew like an immense tree over the other trees. My father occupied the same spot where he had been on Sunday afternoon, but he didn't talk about the rain. He said: "I must have slept poorly last night because I woke with a stiff back." And he stayed there, sitting by the railing with his feet on a chair and his head turned toward the empty garden. Only at dusk, after he had turned down lunch, did he say: "It looks as if it will never clear." And I remembered the months of heat. I remembered August, those long and awesome siestas in which we dropped down to die under the weight of the hour, our clothes sticking to our bodies, hearing outside the insistent and dull buzzing of the hour that never passed. I saw the washed-down walls, the joint of the beams all puffed up by the water. I saw the small garden, empty for the first time, and the jasmine bush against the wall, faithful to the memory of my mother. I saw my father sitting in a rocker, his painful vertebrae resting on a pillow and his sad eyes lost in the labyrinth of the rain. I remembered the August nights in whose wondrous silence nothing could be heard except the millenary sound that the earth makes as it spins on its rusty, unoiled axis. Suddenly I felt overcome by an overwhelming sadness.

It rained all Monday, just like Sunday. But now it seemed to be raining in another way, because something different and bitter was going on in my heart. At dusk a voice beside my chair said: "This rain is a bore." Without turning to look, I recognized Martin's voice. I knew that he was speaking in the next chair, with the same cold and awesome expression that hadn't varied, not even after that gloomy December dawn when he started being my husband. Five months had passed since then. Now I was going to have a child. And Martin was there beside me saying that the rain bored him. "Not a bore," I said. "It seems terribly sad to me, with the empty garden and those poor trees that can't come in from the courtyard." Then I turned to look at him and Martin was no longer there. It was only a voice that was saying to me: "It doesn't look as if it will ever clear," and when I looked toward the voice I found only the empty chair.

On Tuesday morning we found a cow in the garden. It looked like a clay promontory in its hard and rebellious immobility, its hooves sunken in the mud and its head bent over. During the morning the Indians tried to drive it away with sticks and stones. But the cow stayed there, imperturbable in the garden, hard, inviolable, its hooves still sunken in the mud and its huge head humiliated by the rain. The Indians harassed it until my father's patient tolerance came to its defence. "Leave her alone," he said. "She'll leave the way she came."

At sundown on Tuesday the water tightened and hurt, like a shroud over the heart. The coolness of the first morning began to change into a hot and sticky humidity. The temperature was neither cold nor hot; it was the temperature of a fever chill. Feet sweated inside shoes. It was hard to say what was more disagreeable, bare skin or the contact of clothing on skin. All activity had ceased in the house. We sat on the veranda but we no longer felt it was falling. We no longer saw anything except the outline of the trees in the mist, with a sad and desolate sunset which left on your lips the same taste with which you awaken after having dreamed about a stranger. I knew that it was Tuesday and I remembered the twins of Saint Jerome, the blind girls who came to the house every week to sing us simple songs, saddened by the bitter and unprotected prodigy of their voices. Above the rain I heard the blind twins' little song and I imagined them at home, huddling, waiting for the rain to stop so they could go out and sing. The twins of Saint Jerome wouldn't come that day, I thought, nor would the beggar woman be on the veranda after siesta, asking, as on every Tuesday, for the eternal branch of lemon balm.

That day we lost track of meals. At siesta time my stepmother served a plate of tasteless soup and a piece of stale bread. But actually we hadn't eaten since sunset on Monday and I think that from then on we stopped thinking. We were paralysed, drugged by the rain, given over to the collapse of nature with a peaceful and resigned nature. Only the cow was moving in the afternoon. Suddenly a deep noise shook her insides and her hooves sank into the mud with greater force. Then she stood motionless for half an hour, as if she were already dead but could not fall down because the habit of being alive prevented her, the habit of remaining in one position in the rain, until the habit grew weaker than her body. Then she doubled her front legs (her dark and shiny haunches still raised in a last agonized effort) and sank her drooling snout into the mud, finally surrendering to the weight of her own matter in a silent, gradual, and dignified ceremony of total downfall. "She got that far," someone said behind me. And I turned to look and on the threshold I saw the Tuesday beggar woman who had come through the storm to ask for the branch of lemon balm.

Perhaps on Wednesday I might have grown accustomed to that overwhelming atmosphere if on going to the living room I hadn't found the table pushed against the wall, the furniture piled on top of it, and on the other side, on a parapet prepared during the night, trunks and boxes of household utensils. The spectacle produced a terrible feeling of emptiness in me. Something had happened during the night. The house was in disarray; the Guajiro Indians, shirtless and barefoot, with their pants rolled up to their knees, were carrying the furniture into the dining room. In the men's expression, in the very diligence with which they were working, one could see the cruelty of their frustrated rebellion, of their necessary and humiliating inferiority in the rain. I moved without direction, without will. I felt changed into a desolate meadow sown with algae and lichens, with soft, sticky toadstools, fertilized by the repugnant plants of dampness and shadows. I was in the living room contemplating the desert spectacle of the piled-up furniture when I heard my stepmother's voice warning me from her room that I might catch pneumonia. Only then did I realize that the water was up to my ankles, that the house was flooded, the floor covered by a thick surface of viscous, dead water.

On Wednesday noon it still hadn't finished dawning. And before three o'clock in the afternoon night had come on completely, ahead of time and sickly, with the same slow, monotonous, and pitiless rhythm of the rain in the courtyard. It was a premature dusk, soft and lugubrious, growing in the midst of the silence of the Guajiros, who were squatting on the chairs against the walls, defeated and impotent against the disturbance of nature. That was when news began to arrive from outside. No one brought it to the house. It simply arrived, precise, individualized, as if led by the liquid clay that ran through the streets and dragged household items along, things and more things, the leftovers of a remote catastrophe, rubbish and dead animals. Events that took place on Sunday, when the rain was still the announcement of a providential season, took two days to be known at our house. And on Wednesday the news arrived as if impelled by the very inner dynamism of the storm. It was learned then that the church was flooded and its collapse expected. Someone who had no reason to know said that night: "The train hasn't been able to cross the bridge since Monday. It seems that the river carried away the tracks." And it was learned that a sick woman had disappeared from her bed and had been found that afternoon floating in the courtyard.

Terrified, possessed by the fright and the deluge, I sat down in the rocker with my legs tucked up and my eyes fixed on the damp darkness full of hazy foreboding. My stepmother appeared in the doorway with the lamp held high and her head erect. She looked like a family ghost before whom I felt no fear whatever because I myself shared her supernatural condition. She came over to where I was. She still held her head high and the lamp in the air, and she splashed through the water on the veranda. "Now we have to pray," she said. And I noticed her dry and wrinkled face, as if she had just left her tomb or as if she had been made of some substance different from human matter. She was across from me with her rosary in her hand saying: "Now we have to pray. The water broke open the tombs and now the poor dead are floating in the cemetery."

I may have slept a little that night when I awoke with a start because of a sour and penetrating smell like that of decomposing bodies. I gave a strong shake to Martin, who was snoring beside me. "Don't you notice it?" I asked him. And he said: "What?" And I said: "The smell. It must be the dead people floating along the streets." I was terrified by that idea, but Martin turned to the wall and with a husky and sleepy voice said: "That's something you made up. Pregnant women are always imagining things."

At dawn on Thursday the smells stopped, the sense of distance was lost. The notion of time, upset since the day before, disappeared completely. Then there was no Thursday. What should have been Thursday was a physical, jellylike thing that could have been parted with the hands in order to look into Friday. There were no men or women there. My stepmother, my father, the Indians were adipose and improbable bodies that moved in the marsh of winter. My father said to me: "Don't move away from here until you're told what to do," and his voice was distant and indirect and didn't seem to be perceived by the ear but by touch, which was the only sense that remained active.

But my father didn't return: he got lost in the weather. So when night came I called my stepmother to tell her to accompany me to my bedroom. I had a peaceful and serene sleep, which lasted all through the night. On the following day the atmosphere was still the same, colourless, odourless, and without any temperature. As soon as I awoke I jumped into a chair and remained there without moving, because something told me that there was still a region of my consciousness that hadn't awakened completely. Then I heard the train whistle. The prolonged and sad whistle of the train fleeing the storm. *It must have cleared somewhere*, I thought, and a voice behind me seemed to answer my thought. "Where?" it said. "Who's there?" I asked looking. And I saw my stepmother with a long thin arm in the direction of the wall. "It's me," she said. And I asked her: "Can you hear it?" And she said yes, maybe it had cleared on the outskirts and they'd repaired the tracks. Then she gave me a tray with some steaming breakfast. It smelled of garlic sauce and boiled butter. It was a plate of soup. Disconcerted, I asked my stepmother what time it was. And she, calmly, with a voice that tasted of prostrated resignation, said: "It must be around two-thirty. The train isn't late after all this." I said: "Two-thirty! How could I have slept so long!" And she said: "You haven't slept very long. It can't be more than three o'clock." And I, trembling, feeling the plate slip through my fingers: "Two-thirty on Friday," I said. And she, monstrously tranquil: "Two-thirty on Thursday, child. *Still* two-thirty on Thursday."

I don't know how long I was sunken in that somnambulism where the senses lose their value. I only know that after many uncountable hours I heard a voice in the next room. A voice that said: "Now you can roll the bed to this side." It was a tired voice, but not the voice of a sick person, rather that of a convalescent. Then I heard the sound of the bricks in the water. I remained rigid before I realized that I was in a horizontal position. Then I felt the immense emptiness. I felt the wavering and violent silence of the house, the incredible immobility that affected everything. And suddenly I felt my heart turned into a frozen stone. *I'm dead*, I thought. *My God, I'm dead.* I gave a jump in the bed. I shouted: "Ada! Ada!" Martin's unpleasant voice answered me from the other side. "They can't hear you, they're already outside by now." Only then did I realize that it had cleared and that all around us a silence stretched out, a tranquillity, a mysterious and deep beatitude, a perfect state which must have been very much like death. Then footsteps could be heard on the veranda. A clear and completely living voice was heard. Then a cool breeze shook the panel of the door, made the doorknob squeak, and a solid and monumental body, like a ripe fruit, fell deeply into the cistern in the courtyard. Something in the air revealed the presence of an invisible person who was smiling in the darkness. *Good Lord*, I thought then, confused by the mixup in time. *It wouldn't surprise me now if they were coming to call me to go to last Sunday's Mass.*

Sea and
SUNSET

YUKIO MISHIMA

Yukio Mishima was born in Tokyo in 1925.
In November 1970, he and a group from a private army stormed a military headquarters in Tokyo where Mishima read a proclamation and committed seppuku, a ritual suicide.
On the morning of his death, the last volume of his tetralogy,
The Sea of Fertility, was delivered to his publisher.
"A training in words leads to a fresh discovery of reality... {or, more precisely} a rediscovery."

IT WAS LATE SUMMER IN THE NINTH YEAR OF THE Bunei era. It should be added here, since it will be important later, that the ninth year of Bunei is 1272 by Western reckoning.

An elderly man and a boy were climbing the hill known as Sho-jo-ga-take that lies behind the Kenchoji temple in Kamakura. Even during the summer, the man, a handyman at the temple, liked to get the cleaning done in the middle of the day, then, whenever there promised to be a beautiful sunset, to climb to the top of this hill.

The boy was deaf and dumb, and the village children who came to play at the temple always left him out of their games, so the old man, feeling sorry for him, had brought him up there with him.

The old man's name was Anri. He was not very tall, but his eyes were a clear blue. With his large nose and his deep eye sockets, his appearance was different from ordinary people; so the village brats were accustomed to calling him, behind his back, not Anri but "the long-nosed goblin."

There was nothing at all odd about his Japanese, nor had he any perceptible foreign accent. It was between twenty and twenty-five years since he had arrived here with the Zen Master Daigaku who had founded the temple.

The summer sunlight was beginning to slant its rays, and the area around the Shodo that was sheltered from the sun by the hills was already in shadow. The main gate of the temple reached up to the sky as though marking the boundary between day and night. It was the hour when the whole temple compound, with its many clumps of trees, was suddenly overtaken by shade.

Yet the west side of Shojo-ga-take, up which Anri and the boy were climbing, was still bathed in strong sunlight and the cicadas were clamorous in its woods. Amidst the grasses that gave off a summery odour beside the path, a few crimson spider lilies gave notice of the approach of autumn.

Arrived at the summit, the two of them left their sweat to dry in the cooling breeze off the hills.

Below them the sub-temples making up the Kenchoji lay spread out for their inspection: Sairaiin, Dokeiin, Myokoin, Hojuin, Tengenin, Ryuhoin .... Hard by the main gate, a young juniper tree that the founder had brought as a sapling from his home in Sung China was clearly discernible, gathering the late summer sun to itself on its leaves.

Directly below them on the slope of Shojo-ga-take itself, the roof of the Okunoin was visible, and below it again rose the roof of the belfry. Beneath the cave where the Master had practised meditation, a grove of cherry trees which in season would become a sea of blossom formed a shady retreat of rich green foliage. At the foot of the hill, Daigaku Pond announced its presence with a dull glint of water through the trees.

But it was at something other than these features of the scene that Anri was gazing.

It was the sea, glittering in a distant line beyond the rise and fall of Kamakura's hills and valleys. Throughout the summer one could, from here, watch the sun sink into the sea in the vicinity of Inamura-ga-saki.

Near where the dark blue of the horizon touched the sky lay a low line of cumulus clouds. It did not move, but like a morning glory slowly unfurling its petals was drifting open, gradually changing shape. The sky above it was a clear yet somewhat faded blue, and the clouds, though not coloured as yet, seemed brushed with the faintest tinge of peach by a light coming from within.

The state of the sky showed summer and autumn just coming to terms with each other: for high up, way above the horizon, mackerel clouds stretched across it in a band, a fleecy mass that spread a softly mottled pattern over Kamakura's many valleys.

"Why, they're just like a flock of sheep," said Anri in a voice hoarse with age. But the boy, being deaf and dumb, simply sat there on a nearby rock, gazing intently up at him. Anri might have been talking to himself.

The boy could hear nothing — his mind comprehended nothing — yet his limpid gaze suggested an infinite wisdom, so that it seemed that the feeling Anri wanted to convey, if not his words, might be transmitted directly from Anri's clear blue eyes to the boy's.

Which was why Anri spoke exactly as though he were addressing the boy. His words were not the Japanese of which he normally showed such a fluent command. They were French, a French interlarded with the dialect of the mountainous region in the south where he was born. If any of the other brats had heard him, they would have felt that this smoothly tripping, many-vowelled speech came oddly from the lips of a goblin.

"Yes, just like a flock of sheep ...," repeated Anri, with a sigh this time. "I wonder what happened to those sweet little lambs of the Cévennes? By now, I expect, they've had children, and grandchildren, and great-grandchildren, then died themselves ...."

He lowered himself onto a rock, choosing a place where the summer grasses did not shut off the view of the distant sea.

The hillsides were full of singing cicadas.

Turning his azure eyes in the boy's direction, Anri spoke again.

"I'm sure you don't know what I'm saying. But you are different from those village people — you will probably believe me. It may be difficult, even for you, to accept. But hear it, just the same — no one apart from you is likely to take my story seriously ...."

He spoke haltingly. Whenever he got stuck, he made an odd, unfamiliar gesture as though to help summon up the next part of his tale.

"... A long time ago, when I was around the same age as you are now — no, long before I was your age — I was a shepherd in the Cévennes. The Cévennes is the beautiful southern mountain area of France, the domain of the Count of Toulouse, the area south of Mt. Pilat. I don't suppose that means much to you — after all, the people of this country don't even know the name of my native land!

"It was around the year 1212, a time when the Fifth Crusade

had for a while regained control of the Holy Land, only to see it snatched away once more. The French were sunk in grief, their womenfolk in mourning yet again.

"At dusk one day, I was driving my flock home from the pastures and had started to climb a small hill. The sky was almost unnaturally clear. The dog I had with me gave a low growl, dropped his tail, and behaved as though trying to hide behind me.

"Just then, I saw Christ in a shining white robe coming down the hill in my direction. He had a beard, just as he has in pictures, and his face wore a smile of infinite compassion. I prostrated myself. The Lord stretched out a hand and, I seem to remember, touched my hair. Then he said: 'It is thou, Henri, who shalt take back the Holy Land. Thou and other children like thee shall regain Jerusalem from the heathen Turk. Gather thy fellows, in great numbers, and betake thee to Marseilles. There, the waters shall be divided in two that thou mayst pass beyond, and to the Holy Land.'

"... That much I heard, no mistake about it. For the rest, I was in a swoon. The dog licked my face to wake me, and when I came to it was there, right in front of me, peering anxiously into my face in the twilight. My whole body was bathed in sweat.

"When I got home I spoke to no one of what had happened: I felt that no one would believe me.

"It was four or five days later, a day of rain. I was alone in the shepherds' hut. Around dusk, just as before, there came a knock at the door. Going out, I found an aged traveller, who begged some bread of me. I stared intently at him. It was a solemn visage, with a large, high-bridged nose and white whiskers framing the face. The eyes, especially, had an almost frightening clarity. I bade him come in, as it was raining, but he made no answer. Then I saw that, though he had been walking through the rain, his robe was quite dry.

"I stood there awestruck, unable to speak. The old man thanked me for the bread and took his leave. As he went, I heard a voice say quite clearly in my ear: 'Hast thou forgotten the prophesy made a while since? Why dost thou hesitate? Knowst thou not that thou art entrusted with a mission by the Lord?'

"I made to go after the old man. But it was pitch dark all about, the rain was beating down, and the traveller was nowhere to be seen. The sound of the sheep bleating uneasily as they huddled together came to me through the rain ....

"I could not sleep that night.

"The next day, when I went out into the pasture, I finally told the story to a young shepherd of my own age with whom I was particularly friendly. A pious youth, he no sooner heard my tale than he fell to his knees in the clover and paid homage to me.

"Within a week or so, I had collected a following among the shepherds of the neighbourhood. I had no sense of self-importance, but they came forward of their own accord to be my followers.

"Before long, a rumour got about that an eight-year-old prophet had put in an appearance at a place not far from the village where I lived. The infant seer — so the stories said — was preaching and performing miracles; it was even said that he had laid hands on a blind girl and restored her sight.

"My followers and I went there to see for ourselves. We found the prophet playing and laughing merrily amid the other children. I went down on my knees before him, and told him in detail of the revelation.

"The child had a milky white complexion, and golden curls hung down over a forehead in which the blue veins were visible. As I knelt, he checked his laughter and his small mouth twitched two or three times at the corners. But it was not I he was looking at. He was gazing abstractedly at the meadows' undulating skyline.

"So I too looked in the same direction and saw, standing there, a rather tall olive tree. The light was filtering through its upper foliage, so that branches and leaves seemed to be glowing from within.

"A breeze swept across the land. Placing a solemn hand on my shoulder, the child pointed in that direction. And in the higher branches of the tree I saw quite clearly a host of angels, all moving their wings.

" 'You are to go east,' said the child in a voice entirely different in its solemnity from a moment before. 'Go east, and keep going. You had best go to Marseilles, as the revelation told you.'

"Thus the rumours spread. Similar things had been happening all over France. One day, the son of a man killed in the Crusades had taken up the sword his father had bequeathed him and left home. At another place, a child who had been playing by the fountain in the garden abruptly threw down his toys and went off, taking nothing but a piece of bread the maid had given him. When his mother caught and scolded him, he would not listen, but declared that he was going to Marseilles.

"In one village the children, slipping out of bed in the night, gathered in the village square, then set out for they knew not where, singing hymns as they went. When the grown-ups awoke, the village was bereft of all children save those who were still too small to walk.

"I myself was preparing to leave for Marseilles together with a large company of followers when my parents came to take me home, begging me in tears to give up my wild scheme. But my companions were impatient with such lack of faith and drove them off. Those who set out with me numbered, alone, no less than a hundred; in all, several thousand children from all over France and Germany were taking part in this Crusade.

"The journey was not easy. We had barely gone half a day when the youngest and the weakest began to fall by the way. We buried their bodies and wept, and set up small crosses of wood to mark the spots.

"Another company of a hundred children, I heard, strayed unwittingly into an area where the Black Death was rampant, and died, to the last child. One of our own band, a girl, was made delirious by fatigue and flung herself from a cliff.

"Strangely enough, these children as they died invariably had a vision of the Holy Land: a vision, I'm sure, not of the desolate Holy Land of today, but a fertile place where the lilies bloomed in profusion, a plain flowing with milk and honey.... You may wonder how we knew such things, but some of them described their visions as they died, and even when they didn't their look was ecstatic, as though they were confronting vast realms of light....

"Well: we arrived in Marseilles.

"Scores of boys and girls were there before us, waiting in the hope that the waters of the sea would part when we appeared. Our number had already been reduced to a third by then.

"Surrounded by children with faces alight with anticipation, I went down to the harbour. There, amidst the rows of masts, the sailors gazed at us curiously. Reaching the quayside, I prayed. The sun was dazzling on the late afternoon sea. I prayed for a long time. The sea brimmed with water as ever, the waves beat heedless against the wharf.

"But we did not despair: the Lord was surely waiting till we were all assembled.

"A few at a time, the children arrived. All were exhausted, some grievously sick. Day after day we waited in vain; but the waters did not part.

"It was then that a man with an exceedingly pious air approached and gave us alms. Having done so, he offered, hesitantly enough, to take us to Jerusalem in his own ship. Half our number were reluctant to embark with him, but the other half, myself included, boldly went on board.

"The ship did not go to the Holy Land, but pointed its prow to the south and sailed till it reached Alexandria in Egypt. There, in the slave market, every one of us was sold."

Anri was silent for a while, as though recalling once more the anguish of that time.

Up in the sky a resplendent late-summer sunset had begun. The mackerel clouds were all crimson, and there were other clouds too, as though long red and yellow streamers had been drawn across the sky. Out at sea, the heavens were like a fiercely blazing furnace.

The very grass and trees about them shone to a still brighter green in the sky's reflected flames.

By now Anri's words were aimed directly at the sunset, almost as if appealing to it. In the flames over the sea his eyes could make out the sights of his home, and the faces of its people. He could see himself as a boy. He could see the other shepherd boys, his friends. On hot summer days they would push their robes of rough cloth off their shoulders, revealing the rosy nipples on the fair-skinned, boyish chest. The faces of the young comrades who had been killed, or had died, rose up in a host in the sunset glow of the sea. They were bareheaded, yet their flaxen hair shone like helmets of fire.

Even when they survived, the boys and girls were scattered to the four quarters. Not once in all his long life as a slave did Anri come across a face he knew. He was never, in the end, to visit the Jerusalem he had so longed to see.

He became the slave of a Persian merchant. Sold a second time, he went to India. There, he heard rumours of a conquest of the West by Batu, grandson of Genghis Khan; and he wept at the peril facing his homeland.

In those days the Zen Master Daigaku was in India studying Buddhism. By a chance chain of circumstances, Anri was freed through the Master's aid. As a way of showing his gratitude, he determined that he would serve the Master for the remainder of his life. He followed him back to his own country, then, hearing that he was going over to Japan, made bold to request that he be allowed to accompany him there....

Anri was at peace with himself by now. The vain hope of returning home long since relinquished, he was resigned to laying his bones to rest in Japan. He had taken in the Master's teachings well, and never indulged in foolish fantasies of an afterlife or hankered after unseen lands. And yet when sunset coloured the summer sky and the sea was a shining bar of scarlet, his legs seemed to start walking of their own accord, taking him irresistibly to the top of Shojo-ga-take.

He would watch the sunset, and the reflection on the waves; and irresistibly he would find himself recalling the wondrous things that just once — he was sure — had befallen him in the days when his life was still new. Once more, he would rehearse them to himself: the miracle; the yearning for the unknown; the strange force that had driven them to Marseilles. And, last of all, he would think of the sea and how, when he had prayed on the quayside of Marseilles in a crowd of children, it had not parted to let them pass but had gone on sending in its placid waves, glittering beneath the setting sun.

Just when he had lost his faith Anri could not remember. The one thing he could recall, vividly even now, was the mystery of the sea, aglow in the sunset, whose waters had failed to give way however much they prayed: a fact more incomprehensible than any miraculous vision. The mystery of that encounter between a boyish mind that saw nothing strange in a vision of Christ, and a sunset sea that refused absolutely to divide...

If at any time in his life the sea had been going to part, it should have done so at that moment, yet even then it had stretched silent, fiery still in the sunset: there lay the mystery....

Silent now, the old man from the temple stood there, the

light glowing on his dishevelled white hair and planting spots of scarlet in the clear azure of his eyes.

The late summer sun was beginning to sink off the coast of Inamura-ga-saki. The surrounding sea turned to a tide of blood.

He remembered the past, remembered the scenes and the people of his home. But now there was no desire to go back; for all of them — the Cévennes, the sheep, his native land — had vanished into the sunset sea. They had vanished, one and all, when the waves had refused to give way.

Anri, though, kept his eyes on the sunset as it changed colour from moment to moment, consuming itself little by little and turning to ash.

The trees and plants of Shojo-ga-take, finally overtaken by shadow, showed all the more clearly the veins of their leaves and the contours of the knots in the wood. Some of the many lesser temple buildings were already sunk in the dusk.

The shadows were creeping up around Anri's feet; the sky above had drained to a dark, grayish blue. A glitter still lingered far offshore, but it had been squeezed by the dusky sky into a thin strip of gold and vermilion.

Just then, from below Anri as he lingered, there rose the deep boom of a temple bell; the belfry on the slope of the hill had begun to toll the end of day.

The sound of the bell came in slow waves that seemed to awake pulsations in the darkness, spreading it out in all directions. The gravely swaying sound did not so much tell the time as instantly dissolve it and carry it away into eternity.

Anri listened, his eyes closed. When he opened them again, he was submerged in dusk and the sea's distant border showed dim and ashen. The sunset was over.

Anri turned to the boy to suggest that they start back for the temple. The boy sat with arms clasped about his knees, on which his head rested. He was fast asleep.

Journey to

DHARMSALA

ROHINTON MISTRY

Rohinton Mistry was born in Bombay in 1952 and emigrated to Toronto in 1975. His first novel, Such a Long Journey, *won the Governor General's Award in 1991. He reflects, "I suppose we are all listeners. Some of us like to tell stories. Do all people wish they could tell stories? I wonder. I sometimes think there is a latent desire in all of us to be story-tellers. In the best of all possible worlds, all of us would be story-tellers and listeners."*

IT WAS STILL RAINING WHEN WE STOPPED outside Hotel Bhagsu. I took my socks off the taxi's corroded chrome door-handles where they had hung to dry for almost four hours, and pulled them over my clammy feet. The socks were still soggy. Little rivulets ran out of my shoulder bag as I squelched into the lobby. The desk clerk watched with interest while I fastidiously avoided a trail of water that ran from the leaky umbrella stand to the door. Why, with the shoes already sopping wet? he must have wondered. I was not sure myself — perhaps to emphasize that I did not generally go about dripping water.

As I signed the register, shaking raindrops from my hands, the desk clerk said that candles would be sent to my room before dark. Candles, I asked?

He had assumed I would know: "There is a small problem. Electricity workers are on strike." Worse, the strikers were sabotaging the power lines. No electricity anywhere, he emphasized, in case I was considering another hotel: not in Upper Dharmsala, not in Lower Dharmsala, nowhere in Kangra District.

I nodded, putting out my hand for the room key. But he held on to it. With that circular motion of the head which can mean almost anything, he said, "There is one more problem." He continued after a suitable pause: "There is also no water. Because of heavy rains. Rocks fell from the mountains and broke all of the water pipes."

He seemed surprised by the lack of emotion with which I greeted his news. But I had already glimpsed the handiwork of the pipe-breaking avalanches during my four-hour taxi ride. The car had laboured hard to reach McLeod Gunj, up the winding, rock-strewn mountain roads, grinding gears painfully, screeching and wheezing, negotiating segments that had become all but impassable.

Perhaps a bit disappointed by my stolidity, once again the desk clerk assured me it was the same in Upper and Lower Dharmsala, and in all of Kangra District; but management would supply two buckets of water a day.

So there was no choice, the hotel would have to do. I requested the day's quota hot, as soon as possible, for a bath. He relinquished my room key at last. Its brass tag had Hotel Bhagsu engraved on one side. "What is Bhagsu?" I asked him, picking up my bag.

"In local language, means Running Water," he said.

The room had an enormous picture window. The curtains, when thrown open, revealed a spectacular view of Kangra Valley. But I could not linger long over it, urgent matters were at hand. I unzipped the bag and wrung out my clothes, spreading them everywhere: over the bed, the chair, the desk, the door knob. Wet and wretched, I sat shivering on the edge of the bed, waiting for the hot water and remembering the warnings to stay away from Dharmsala while it was in the clutches of the dreaded monsoon.

When the Dalai Lama fled Tibet in 1959, just hours before the Chinese conducted a murderous raid on his Palace in Lhasa and occupied the country, he found refuge in India. For months afterwards other Tibetans followed him, anxious to be with their beloved spiritual leader. The pathetic bands of refugees arrived, starving and frostbitten — the ones lucky enough to survive the gauntlet of treacherous mountain passes, the killing cold, and, of course Chou En-lai's soldiers. Each arriving group narrated events more horrific than the previous one: how the Chinese had pillaged the monasteries, crucified the Buddhist monks, forced nuns to publicly copulate with monks before executing them, and were now systematically engaged in wiping out all traces of Tibetan culture.

The Dalai Lama (whose many wonderfully lyrical, euphonious names include Precious Protector, Gentle Glory, and Ocean of Wisdom) spent his first months of exile in anguish and uncertainty. Faced with unabating news of the endless atrocities upon the body and soul of Tibet, he eventually decided that Dharmsala was where he would establish a government-in-exile. Perhaps this quiet mountain hamlet in the Himalayas reminded him of his own land of ice and snow. Soon, a Tibetan colony evolved in Dharmsala, a virtual country-within-a-country. Visitors began arriving from all over the world to see Namgyal Monastery, Tibetan Children's Village, the Dalai Lama's new temple, or to study at the Library of Tibetan Works and Archives.

As a child, it always struck me with wonderment and incredulity that I should have an uncle who lived in Dharmsala. In this remote mountain hamlet he ran the business which has been in the Nowrojee family for five generations. To me, a thousand miles away in Bombay, this land of mountains and snow had seemed miraculously foreign. Photographs would arrive from time to time, of uncle and aunt and cousins wrapped in heavy woollens, standing beside three-foot-deep snow drifts outside their homes, the snow on the roof like thick icing on a cake, and the tree branches delicately lined with more of the glorious white substance. And in my hot and sticky coastal city, gazing with longing and fascination at the photographs, I would find it difficult to believe that such a magical place could exist in this torrid country. Now there, somewhere in the mountains, was a place of escape from heat and dust and grime. So, to visit Dharmsala became the dream.

But for one reason or another, the trip was never taken. Those old photographs: snow-covered mountains and mountain trails; my cousins playing with their huge black Labrador; uncle and aunt posing in the *gaddi* dress of native hill people, a large hookah between them: those old black-and-white photographs curled and faded to brown and yellow. Years passed, the dog died, my cousins got married and settled elsewhere, and my uncle and aunt grew old. Somehow, the thousand miles between Bombay and Dharmsala were never covered. There was always some logistical or financial problem, and travelling third class on Indian trains was only for the foolish or the desperate.

Then, by a quirk of fate, I undertook a different journey ten thousand miles long, to Canada, and I often thought about the irony of it. So this time, back in Bombay to visit family and friends, not monsoon rain nor ticket queues nor diarrhea nor avalanche could keep me away from Dharmsala.

Thus twenty-eight hours by train (first class) brought me to Chakki Bank, in Punjab. It was pouring relentlessly as the first leg of the long journey ended. "Rickshaw, *seth*, rickshaw?" said a voice as I stepped off the train. I quickly calculated: there could be a big demand for transportation in this weather, it might be prudent to say yes. "Yes," I said, and settled the price to Pathankot bus station.

Outside, auto rickshaws — three wheelers — were parked along the station building in a long line. Enough for everyone, I thought. They had black vinyl tops, and plastic flaps at the side which could be fastened shut, I noted approvingly. I followed my man.

And we came to the end of the line. There, he placed my bag in a pitiful cycle rickshaw, the only one amidst that reassuringly formidable squadron of auto rickshaws. The cycle rickshaw had open sides; and old gunny sacks tied to the top of the frame formed a feeble canopy. I watched in disbelief, appalled by my bad luck. No, stupidity, I corrected myself, for it was clear now why he had come inside the station to solicit a fare. That should have made me suspicious. Once upon a time it would have.

The cycle rickshawalla saw my reaction. He pointed pleadingly at the seat, and I looked him in the face, something I never should have done. I am trusting you, his eyes said, not to break our contract. The auto rickshaws taunted me with their waterproof interiors as I stared longingly after them. Their owners were watching, amused, certain I would cave in. And that settled it for me.

Within seconds of setting off, I was rueing my pride. The gunny sacks were as effective as a broken sieve in keeping out the rain, and despite my raincoat I was soon drenched. The downpour saturated my bag and its contents — I could almost feel its weight increasing, minute by minute. The cycle rickshawalla struggled to pedal as fast as he could through streets ankle-deep in water. His calf muscles contracted and rippled, knotting with the strain, and a mixture of pity and anger confused me. I wished the ride would end quickly.

In Pathankot, he convinced me a taxi was better than a bus in this weather. Afterwards, I was glad I took his advice: on the mountains, buses had pulled over because the avalanches, the pipe-breaking avalanches, had made the roads far too narrow. Meanwhile, I waited as the rickshawalla and the taxi driver haggled over the former's commission.

And four hours later I was draping my underwear, socks, shirts and pants over the door knob, armchair, lampshade and window. There was a knock. The houseboy (who doubled as waiter, I discovered later in the restaurant) staggered in with two steaming plastic buckets, one red and the other blue. He looked around disbelievingly at my impromptu haberdashery. "All wet," I explained. He smiled and nodded to humour the eccentric occupant.

I wondered briefly where the water in the buckets came from if the pipes were broken. My guess was a well. In the bathroom, I splashed the hot water over me with a mug.

Dharmsala is a collection of settlements perched across the lower ridges of the Dhauladur range. The Dhauladur range itself is a southern spur of the Himalayas, and surrounds the Kangra Valley like a snow-capped fence. McLeod Gunj, at seven thousand feet, is one of the highest settlements. I had passed others on my way up by taxi: Lower Dharmsala and Kotwali Bazaar, the main commercial centre crowded with hotels, shops, and restaurants; Forsyth Gunj, a one-street village; and, of course, the huge military cantonment, which was the beginning of everything, back in the British days.

Early in this century, the British were considering making Dharmsala their summer capital; they found the plains unbearable in the hot season. But an earthquake badly damaged the place in 1905, and they chose another hill station, Simla, a bit further south. (Later, my uncle would describe it differently: the official in charge of selecting the capital was travelling from Dalhousie to Dharmsala when he caught dysentery on the way, reached Dharmsala and died. The idea of Dharmsala as summer capital was promptly abandoned.)

I wanted to see more of McLeod Gunj and Upper Dharmsala. But first I was anxious to meet my aunt and uncle. Next morning, I telephoned them at their general store, and they were delighted to hear my voice. The line was so bad, they thought I was calling from Bombay. No, I said, Hotel Bhagsu, and they insisted I come immediately, their place was only a five-minute walk away.

It was still drizzling. Along the side of the hotel, under every rain spout, was a plastic bucket. My red and blue were there as well. The houseboy was standing guard over them, watching them fill with the runoff from the roof. He looked away guiltily at first when he saw me. Then he must have decided to put the best face on things, for he acknowledged me by smiling and waving. He seemed like a child caught red-handed at mischief.

My uncle and aunt were very sorry for the way my visit had begun. "But didn't anyone tell you? This is not a good season for Dharmsala," they said. I had been warned, I admitted, but decided to come anyway. They found this touching, and also confusing. Never mind, uncle said, perhaps half our troubles would soon be over: the military cantonment had dispatched its men to find and repair the sabotaged power lines. The only snag was, as soon as they mended one, the strikers snipped through some more.

As for water, said my aunt, not to worry, their supply had not been affected, I could shower here.

Not affected? How? Just then, customers arrived, asking for candles. My aunt went to serve them and my uncle told the story.

During the devastation of the 1905 earthquake, the Nowrojee Store was practically the only structure that survived. Uncle's grandfather had handed out food and clothing and blankets from store supplies till proper relief was organized by the British District Commissioner. When McLeod Gunj was back on its feet, the District Commissioner wanted to show his gratitude to the family. He gifted a mountain spring to them, and arranged for direct water supply from the spring to their house. That private pipeline was still operating after eighty-odd years, and had survived the present avalanches.

I promised I would use their shower in the evening. Then more customers entered, and he had to assist my aunt. Local people were inquiring if the newspaper delivery was expected to get through to Dharmsala. Foreign tourists in designer raincoats were seeking out the sturdy black umbrella which, locally, was the staple defence against the rains. The tourists were also laying in a stock of Bisleri mineral water.

There was a lull in business after this surge. My aunt suggested that uncle take me around Dharmsala for a bit, she could hold the fort alone. So we set off for a walk.

At first the going was slow. Almost every person we passed

stopped to exchange a few words, mainly about the weather, and which roads were closed and which were still passable. But it was heartening to see the Tibetan monks, in their crimson robes, always smiling joyfully. For a people who had suffered such hardships and upheavals, struggling to start life over again in a strange land, they were remarkably cheerful and happy. Perhaps this, and their Buddhist faith, is what sustained them. They had the most wonderful beaming, smiling faces. Just like their spiritual leader, whom I had watched some time ago on "Sixty Minutes," whose countenance seems to radiate an inner well-being.

Exchanging *namaskaars* with everyone we met (the folded-hands greeting, which translates into: I greet the God in you, common to Hindus and Buddhists), we arrived at a tall gold-crowned structure at the centre of a group of buildings. It was a *chorten,* a religious monument, dedicated to the memory of all those suffering under Chinese occupation in Tibet. The faithful were circling round it, spinning two rows of prayer wheels and reciting mantras.

We left the little square and the buildings which housed Tibetan handicraft shops, restaurants and hotels. Further down were the Tibetan homes: shacks and shanties of tin and stone, and every window was adorned with flowers in rusty tin cans. Faded prayer flags fluttered in the trees overhead.

The road climbed steeply. Before I knew it, the buildings and the chorten were below us. My uncle turned and pointed. There used to be a beautiful park there, he said, at the centre of McLeod Gunj, but it had to go when the refugees came.

During our walk I had gathered he loved the Tibetan people, and had done much to aid them. I could hear the respect and admiration in his voice when he talked about the Dalai Lama, whom he had helped, back in 1959, to acquire suitable houses and properties where the Tibetans could start rebuilding their lives. But now as my uncle told the story of Dharmsala and the arrival of the refugees, I could not help feeling that there was also some resentment towards these people who had so radically changed and remade in their own image the place where he was born, the place he loved so dearly. My aunt, who likes the hustle and bustle of big cities and gets her share of it by visiting relatives periodically, said he would pine away if she ever insisted they leave Dharmsala.

We continued to climb, and on the mountain spur that dominates the valley rose the golden pinnacles of Thekchen Choeling, the Island of Mahayan Teaching, the complex which was the new residence of the Dalai Lama. His cottage had a green corrugated roof, and the temple was a three-storey lemon-yellow hall topped by gold spires. On a low verandah surrounding the temple, a woman was performing repeated prostrations. She was making a circuit of the temple, measuring her progress with her height.

We removed our shoes and went inside. The main hall had a high throne at one end: the Dalai Lama's throne, on which he sat when he gave audiences and preached. There would be no audiences for the next few days, though, because he was away in Ladakh to deliver the Kalachakra — Wheel of Time — Initiation. Behind the throne there was a larger-than-life statue of the Buddha in the lotus position. The Buddha was locked in a huge glass case. Myriads of precious and semi-precious stones formed a halo around the Buddha's solid gold head, and hence the locked glass: things had changed in Dharmsala; the increase in population and the tourist traffic forced the monks to take precautions.

The changes were having other effects, too. The mountain slopes were being rapidly deforested by the poverty-stricken population's hunger for firewood. And, as elsewhere in the world, the disappearance of trees was followed by soil erosion. My uncle had pointed out the gashed and scarred hills on our climb. He said that so many mudslides and rockfalls were unheard of in the old days; and there was less and less snow each year.

I thought of those photographs from my childhood. Their memory suddenly seemed more precious than ever. The pristine place they had once captured was disappearing,

Inside the temple, at the throne's right, more statues were displayed. One of them had multiple heads and arms: Chenrezi, the awareness-being who symbolizes compassion in the Tibetan pantheon. The legend went that Chenrezi was contemplating how best to work for the happiness of all living things, when his head burst into a thousand pieces as he realized the awesome nature of the task. The Buddha of Limitless Light restored him to life, giving him a thousand heads to represent the all-seeing nature of his compassion, and a thousand arms to symbolize the omnipresence of his help. But now Chenrezi, along with other statues bedecked with gold and jewels, was locked behind a floor-to-ceiling collapsible steel gate.

The rain finally ceased. My uncle wished the mist would clear so he could show me Pong Lake in the distance. When the moon shone upon the water, he said, it took one's breath away. But the mist sat over the valley, unmoving.

Descending the temple road, we saw several monks, prayer beads in hand, walking a circular path around the complex. They were simulating the Lingkhor, the Holy Walk circumscribing the Potala, the Dalai Lama's palace in Tibet. Round and round they walked, praying, perhaps, for a time when he would be back in his palace, and they treading the original Lingkhor.

Inside: the woman, making a mandala of her prostrations around the temple. Outside: the monks, creating circles of prayer around their beloved leader's residence. Circles within circles. The Wheel of Time.

Back at the general store, bad news awaited: the taps were dry. The Tibetan refugees (everyone, Tibetans included, used that word, despite their having lived here thirty years; perhaps clinging to this word kept alive the hope of returning to their Land of Snows) had discovered that the Nowrojee pipeline still held water. They had cut it open to fill their buckets. Strangely, my uncle and aunt were not too upset. It had happened before. They just wished the people would come to the house and fill their buckets from the taps instead of cutting the pipes.

Later that night, I found my way back to Hotel Bhagsu with a borrowed flashlight. My uncle accompanied me part of the way. Near the incline that led to the hotel, where the road forked, there was a little lamp in an earthen pot, sitting at the very point of divergence. How quaint, I thought. A friendly light to guide the traveller through the pitch-black night. He said to tread carefully to the right of the lamp, by no means to step over it.

What was it? Something to do with Tibetan exorcism rites, he answered. Did he believe in such things? He had lived here too long, he said, and seen too much, to be able to disbelieve it completely. Despite my scepticism, he succeeded in sending a shiver down my spine. It was only the setting, I explained to myself: a pitch-dark mountain road, the rustling of leaves, swirling mists.

Back at the hotel, the desk clerk apologetically handed me the stubs of two candles. Dharmsala was out of candles, what remained had to be strictly rationed. I asked for water.

One more day, I decided, then I would leave. There was not much to do. The avalanches had closed the roads further north, and the side trips I had planned to Dalhousie, Kulu, and Manali were not feasible. The houseboy knocked.

He was carrying the red bucket. "Where is the blue?" He shook his head: "Sorry, not enough rain. Today only one bucket."

The electricity was back next morning, I discovered thankfully. Around nine, I went to the empty restaurant and ordered tea and toast. Afflicted with a bad stomach, I had been virtually living on toast for the past three days. The houseboy in the persona of waiter took my order cheerfully and left.

Thirty minutes later I was still waiting. The door marked Employees Only was ajar, and I peered into the kitchen. It was empty. The backyard beyond the kitchen window was deserted too. I went to the front desk. No one. Finally, I ran into the night watchman who had just woken. "What is going on?" I asked him with manufactured testiness, remembering long-forgotten roles and poses. "Waiter has disappeared, no one in the kitchen, no one on duty. What has happened? Is this a hotel or a joke?"

He studied his watch and thought for a moment: "Sunday today? Oh yes. Everyone is watching *Ramayan.* But they will come back. Only five minutes left."

The Ramayana is one of the two great Sanskrit epics of ancient India. The other is Mahabharata, which recently found its way in translation onto Western stages in Peter Brook's production. But when the Ramayana, the story of the god Rama, was made into a Hindi TV serial, sixty million homes began tuning in every Sunday morning, and those who did not own TVs went to friends who did. In the countryside, entire villages gathered around the community set. Before the program started, people would garland the TV with fresh flowers and burn incense beside it. Classified ads in newspapers would read: Car For Sale — But Call After *Ramayan.* Interstate buses would make unscheduled stops when the auspicious time neared, and woe betide the bus driver who refused. Ministerial swearing-in ceremonies were also known to be postponed.

The series ended after seventy-eight episodes which, however, were not sufficient to cover the entire epic. In protest, street sweepers went on strike and there were demonstrations in several cities. The Ministry of Information and Broadcasting then sanctioned a further twenty-six episodes in order to bring *Ramayan* and the strike to their proper conclusions.

But the story does not end there. Not satisfied with burning incense and garlanding their television sets on Sunday mornings, people began mobbing the actor who played the role of Rama, genuflecting wherever he appeared in public, touching his feet, asking for his blessing. To capitalize on the phenomenon, Rajiv Gandhi's Congress Party enlisted the actor-god to campaign for their candidate in the upcoming election. The actor-god went around telling people that Rama would give them blessings if they voted for the Congress Party, and how it was the one sure way to usher in the golden age of Rama's mythical kingdom of Ayodhya.

At this point, the intellectuals and political pundits sadly shook their sage heads, lamenting the ill-prepared state of the masses for democracy. Suspension of disbelief was all very well when watching television. But to extend it to real life? It showed, they said, the need for education as a prerequisite if democracy was to work successfully.

When it was time to vote, however, the masses, despite the actor-god and the shaking heads of the intellectuals, knew exactly what to do. The Congress candidate went down in a resounding defeat, and the actor-god became sadly human again.

My waiter returned, promising immediate delivery of my tea and toast. I threw my hands in the air and pretended to be upset: How long was a person supposed to wait? In response to my spurious annoyance, he affected a contrite look. But, like me, his heart was not in it. Like the voters and the actor-god, we played out our roles, and we both knew what was what.

In Bombay, at the beginning of the trip, I had listened amusedly when told about the power of the serial. Intriguing me was the fact that what was, by all accounts, a barely passable production lacking any kind of depth, with embarrassingly wooden acting, could, for seventy-eight weeks, hold a captive audience made up of not only Hindus but also Muslims, Sikhs, Parsis, Christians —

cutting right across the religious spectrum. Could it be that under the pernicious currents of communalism and prejudice there were traces of something more significant, a yearning, perhaps, which transcended these nasty things, so that the great Sanskrit epic of ancient India, a national heritage, could belong to all Indians?

I had not expected to receive a personal demonstration of the Sunday morning power that *Ramayan* wielded. Least of all in this faraway mountain hamlet. In a way, though, it was fitting. Everywhere, *Ramayan* brought diverse communities together for a short while, to share an experience. But in Dharmsala, the native population and the refugees have been sharing and living together for many years. Even the electricity saboteurs cooperated with the show. Of course, shortly after *Ramayan* the region was once again powerless.

Halfway between McLeod Gunj and Forsyth Gunj was an old English church my uncle had told me about. The pure scent of pine was in the air as I walked to it. The rock face of the mountain appeared to have burst into fresh green overnight. The rains had given birth to countless little streams and rivulets that gurgled their descent. Sometimes, at a bend in the road, the noise of the water was so loud, it seemed that a huge waterfall was waiting round the corner. But it was only the wind and the mountains playing tricks, orchestrating, weaving and blending the music of the newborn runnels and brooks into one mighty symphony of a cataract.

The church of the beautiful name came into view: the Church of Saint John in the Wilderness, a lonely reminder of the British Raj. It looked very much like any English parish church. The grounds were in grave neglect. A tall pine had fallen across the walkway, brought down by the rains, no doubt. Sunday morning service was in progress. Tourists and local residents made up the scanty congregation.

I walked around to the back and found myself in the churchyard. A ten- or twelve-foot monument dominated the cemetery. Intrigued, I went closer. James Bruce, Eighth Earl of Elgin and Twelfth Earl of Kincardine (1811-1863), read the inscription, barely legible. And then, the positions he had held in the far-flung corners of the Empire: Governor of Jamaica, Governor General of Canada, Viceroy and Governor General of India.

I examined other gravestones. But the weather and time had

successfully effaced most of the words. A date here, a first name there
— Dear Wife ..., or: Faithful Husband ..., and then, Final rest ...,
and: Heavenly Peace — these fragments were all I could read.

I went back to Lord Elgin's grave and sat before it on a stone
ledge. The churchyard was deserted. I read again the words carved
in stone, thinking about this Viceroy who had died in Dharmsala, so
far from his own country. I imagined the long journeys he had
undertaken for Queen and Country: what had he thought about this
ancient country? Had he enjoyed his stay here? How might he have
felt at having to live out his life in distant lands, none of them his
home? Sitting on the moss-grown ledge, I thought about this man
buried here, who, one hundred and twenty-five years ago and more,
had governed them both, my old country and my new; I thought
about the final things.

The weather-beaten gravestones, the vanished epitaphs, the
disappearing inscriptions, somehow brought back to me the fading,
indistinct photographs of uncle and aunt, cousins and dog, snow on
the rooftops and trees. How far away was it — that Dharmsala of my
imagination and of my uncle's youth — how far from what I'd seen?
As far away, perhaps, as the world of empire in whose cause Lord
Elgin had undertaken his travels.

I thought about my own journey: from the Dharmsala of
childhood fantasies to the peaceful churchyard of Saint John in the
Wilderness; and then, amidst the gentle ruins of weatherworn,
crumbling gravestones, back to the fading, curling photographs. To
have made this journey, I felt, was to have described a circle of my
own. And this understanding increased the serenity of the moment.

It started to drizzle. I put on my raincoat and opened my
umbrella. As it gathered strength, the rain streamed down the sides
of Lord Elgin's monument and blurred the words I had been reading.
Thoughts of departure, of descending from the tranquillity of the
mountains into the dusty, frenetic plains, began gnawing at the edges
of the moment. But I pushed them away. I sat there a little longer,
listening to the soothing patter as the rain fell upon the leaves and on
the gravestones all around.

MENESETEUNG

ALICE MUNRO

I

Columbine, bloodroot,
And wild bergamot,
Gathering armfuls,
Giddily we go.

OFFERINGS, THE BOOK IS CALLED. GOLD lettering on a dull-blue cover. The author's full name underneath: Almeda Joynt Roth. The local paper, the *Vidette,* referred to her as "our poetess." There seems to be a mixture of respect and contempt, both for her calling and for her sex — or for their predictable conjuncture. In the front of the book is a photograph, with the photographer's name in one corner, and the date: 1865. The book was published later, in 1873.

The poetess has a long face; a rather long nose; full, sombre dark eyes, which seem ready to roll down her cheeks like giant tears; a lot of dark hair gathered around her face in droopy rolls and curtains. A streak of grey hair plain to see, although she is, in this picture, only twenty-five. Not a pretty girl but the sort of woman who may age well, who probably won't get fat. She wears a tucked and braid-trimmed dark dress or jacket, with a lacy, floppy arrangement of white material — frills or a bow — filling the deep V at the neck. She also wears a hat, which might be made of velvet, in a dark colour to match the dress. It's the untrimmed, shapeless hat, something like a soft beret, that makes me see artistic intentions, or at least a shy and stubborn eccentricity, in this young woman, whose long neck and forward-inclining head indicate as well that she is tall and slender and somewhat awkward. From the waist up, she looks like a young nobleman of another century. But perhaps it was the fashion.

"In 1854," she writes in the preface to her book, "my father brought us — my mother, my sister Catherine, my brother William, and me — to the wilds of Canada West (as it then was). My father was a harness-maker by trade, but a cultivated man who could quote by heart from the Bible, Shakespeare, and the writings of Edmund Burke. He prospered in this newly opened land and was able to set up a harness and leather-goods store, and after a year to build the comfortable house in which I live (alone) today. I was fourteen years old, the eldest of the children, when we came into this country from Kingston, a town whose handsome streets I have not seen again but often remember. My sister was eleven and my brother nine. The third summer that we lived here, my brother and sister were taken ill of a prevalent fever and died within a few days of each other. My dear mother did not regain her spirits after this blow to our family. Her health declined, and after another three years she died. I then became housekeeper to my father and was happy to make his home for twelve years, until he died suddenly one morning at his shop.

"From my earliest years I have delighted in verse and I have occupied myself — sometimes allayed my griefs, which have been no more, I know, than any sojourner on earth must encounter – with many floundering efforts at its composition. My fingers, indeed, were always too clumsy for crochetwork, and those dazzling productions of embroidery which one sees often today — the overflowing fruit and flower baskets, the little Dutch boys, the bonneted maidens with their watering cans — have likewise proved to be beyond my skill. So I offer instead, as the product of my leisure hours, these rude posies, these ballads, couplets, of reflections."

Titles of some of the poems: "Children at Their Games," "The Gypsy Fair," "A Visit to My Family," "Angels in the Snow," "Champlain at the Mouth of the Meneseteung," "The Passing of the Old Forest," and "A Garden Medley." There are some other, shorter poems, about birds and wildflowers and snowstorms. There is some comically intentioned doggerel about what people are thinking about as they listen to the sermon in church.

"Children at Their Games": The writer, a child, is playing

with her brother and sister — one of those games in which children on different sides try to entice and catch each other. She plays on in the deepening twilight, until she realizes that she is alone, and much older. Still she hears the (ghostly) voices of her brother and sister calling. *Come over, come over, let Meda come over.* (Perhaps Almeda was called Meda in the family, or perhaps she shortened her name to fit the poem.)

"The Gypsy Fair": The Gypsies have an encampment near the town, a "fair," where they sell cloth and trinkets, and the writer as a child is afraid that she may be stolen by them, taken away from her family. Instead, her family has been taken away from her, stolen by Gypsies she can't locate or bargain with.

"A Visit to My Family": A visit to the cemetery, a one-sided conversation.

"Angels in the Snow": The writer once taught her brother and sister to make "angels" by lying down in the snow and moving their arms to create wing shapes. Her brother always jumped up carelessly, leaving an angel with a crippled wing. Will this be made perfect in Heaven, or will he be flying with his own makeshift, in circles?

"Champlain at the Mouth of the Meneseteung": This poem celebrates the popular, untrue belief that the explorer sailed down the eastern shore of Lake Huron and landed at the mouth of the major river.

"The Passing of the Old Forest": A list of all the trees — their names, appearance, and uses — that were cut down in the original forest, with a general description of the bears, wolves, eagles, deer, waterfowl.

"A Garden Medley": Perhaps planned as a companion to the forest poem. Catalogue of plants brought from European countries, with bits of history and legend attached, and final Canadianness resulting from this mixture.

The poems are written in quatrains or couplets. There are a couple of attempts at sonnets, but mostly the rhyme scheme is simple — *abab* or *abcb*. The rhyme used is what was once called "masculine" ("shore"/"before"), though once in a while it is "feminine" ("quiver"/"river"). Are those terms familiar anymore? No poem is unrhymed.

II

White roses cold as snow
Bloom where those "angels" lie.
Do they but rest below
Or, in God's wonder, fly?

In 1879, Almeda Roth was still living in the house at the corner of
Pearl and Dufferin streets, the house her father had built for his
family. The house is there today: the manager of the liquor store
lives in it. It's covered with aluminum siding; a closed-in porch has
replaced the veranda. The woodshed, the fence, the gates, the privy,
the barn — all these are gone. A photograph taken in the eighteen-
eighties shows them all in place. The house and fence look a little
shabby, in need of paint, but perhaps that is just because of the
bleached-out look of the brownish photograph. The lace-curtained
windows look like white eyes. No big shade tree is in sight, and, in
fact, the tall elms that overshadowed the town until the nineteen-
fifties, as well as the maples that shade it now, are skinny young trees
with rough fences around them to protect them from the cows.
Without the shelter of those trees, there is a great exposure — back
yards, clotheslines, woodpiles, patchy sheds and barns and privies —
all bare, exposed, provisional looking. Few houses would have
anything like a lawn, just a patch of plantains and anthills and raked
dirt. Perhaps petunias growing on top of a stump, in a round box.
Only the main street is gravelled; the other streets are dirt roads,
muddy or dusty according to season. Yards must be fenced to keep
animals out. Cows are tethered in vacant lots or pastured in back
yards, but sometimes they get loose. Pigs get loose, too, and dogs
roam free or nap in a lordly way on the boardwalks. The town has
taken root, it's not going to vanish, yet it still has some of the look of
an encampment. And, like an encampment, it's busy all the time —
full of people who, within the town, usually walk wherever they're
going; full of animals, which leave horse buns, cowpats, dog turds,
that ladies have to hitch up their skirts for; full of the noise of
building and of drivers shouting at their horses and of the trains that
come in several times a day.

I read about that life in the *Vidette*.

The population is younger than it is now, than it will ever be again. People past fifty usually don't come to a raw, new place. There are quite a few people in the cemetery already, but most of them died young, in accidents or childbirth or epidemics. It's youth that's in evidence in town. Children — boys — rove through the streets in gangs. School is compulsory for only four months a year, and there are lots of occasional jobs that even a child of eight or nine can do — pulling flax, holding horses, delivering groceries, sweeping the boardwalk in front of stores. A good deal of time they spend looking for adventures, One day they follow an old woman, a drunk nicknamed Queen Aggie. They get her into a wheelbarrow and trundle her all over town, then dump her into a ditch to sober her up. They also spend a lot of time around the railway station. They jump on shunting cars and dart between them and dare each other to take chances, which once in a while result in their getting maimed or killed. And they keep an eye out for any strangers coming into town. They follow them, offer to carry their bags, and direct them (for a five-cent piece) to a hotel. Strangers who don't look so prosperous are taunted and tormented. Speculation surrounds all of them — it's like a cloud of flies. Are they coming to town to start up a new business, to persuade people to invest in some scheme, to sell cures or gimmicks, to preach on the street corners? All these things are possible any day of the week. Be on your guard, the *Vidette* tells people. These are times of opportunity and danger. Tramps, confidence men, hucksters, shysters, plain thieves, are travelling the roads, and particularly the railroads. Thefts are announced: money invested and never seen again, a pair of trousers taken from the clothesline, wood from the woodpile, eggs from the henhouse. Such incidents increase in the hot weather.

Hot weather brings accidents, too. More horses run wild then, upsetting buggies. Hands caught in the wringer while doing the washing, a man lopped in two at the sawmill, a leaping boy killed in a fall of lumber at the lumberyard. Nobody sleeps well. Babies wither with summer complaint, and fat people can't catch their breath. Bodies must be buried in a hurry. One day a man goes through the streets ringing a cowbell and calling "Repent! Repent!" It's not a stranger this time, it's a young man who works at the butcher shop. Take him home, wrap him in cold wet cloths, give him some nerve medicine, keep him in bed, pray for his wits. If he doesn't recover, he must go to the asylum.

Almeda Roth's house faces on Dufferin Street, which is a street of considerable respectability. On this street merchants, a mill owner, an operator of salt wells, have their houses. But Pearl Street, which her back windows overlook and her back gate opens onto, is another story. Workmen's houses are adjacent to hers. Small but decent row houses — that is all right. Things deteriorate toward the end of the block, and the next, last one becomes dismal. Nobody but the poorest people, the unrespectable and undeserving poor, would live there at the edge of a boghole (drained since then), called the Pearl Street Swamp. Bushy and luxuriant weeds grow there, makeshift shacks have been put up, there are piles of refuse and debris and crowds of runty children, slops are flung from doorways. The town tries to compel these people to build privies, but they would just as soon go in the bushes. If a gang of boys goes down there in search of adventure, it's likely they'll get more than they bargained for. It is said that even the town constable won't go down Pearl Street on a Saturday night. Almeda Roth has never walked past the row housing. In one of those houses lives the young girl Annie, who helps her with her housecleaning. That young girl herself, being a decent girl, has never walked down to the last block or the swamp. No decent woman ever would.

But that same swamp, lying to the east of Almeda Roth's house, presents a fine sight at dawn. Almeda sleeps at the back of the house. She keeps to the same bedroom she once shared with her sister Catherine — she would not think of moving to the larger front bedroom, where her mother used to lie in bed all day, and which was later the solitary domain of her father. From her window she can see the sun rising, the swamp mist filling with light, the bulky, nearest trees floating against that mist and the trees behind turning transparent. Swamp oaks, soft maples, tamarack, bitternut.

III

Here where the river meets the inland sea.
Spreading her blue skirts from the solemn wood,
I think of birds and beasts and vanished men,
Whose pointed dwellings on these pale sands stood.

One of the strangers who arrived at the railway station a few years ago was Jarvis Poulter, who now occupies the house next to Almeda Roth's — separated from hers by a vacant lot, which he has bought, on Dufferin street. The house is plainer than the Roth house and has no fruit trees or flowers planted around it. It is understood that this is a natural result of Jarvis Poulter's being a widower and living alone. A man may keep his house decent, but he will never — if he is a proper man — do much to decorate it. Marriage forces him to live with more ornament as well as sentiment, and it protects him, also, from the extremities of his own nature — from a frigid parsimony or a luxuriant sloth, from squalor, and from excessive sleeping, drinking, smoking, or freethinking.

> In the interests of the economy, it is believed, a certain estimable gentleman of our town persists in fetching water from the public tap and supplementing his fuel supply by picking up the loose coal along the railway track. Does he think to repay the town or the railway company with a supply of free salt?

This is the *Vidette,* full of shy jokes, innuendo, plain accusation, that no newspaper would get away with today. It's Jarvis Poulter they're talking about — though in other passages he is spoken of with great respect, as a civil magistrate, an employer, a churchman. He is close, that's all. An eccentric, to a degree. All of which may be a result of his single condition, his widower's life. Even carrying his water from the town tap and filling his coal pail along the railway track. This is a decent citizen, prosperous: a tall — slightly paunchy? — man in a dark suit with polished boots. A beard? Black hair streaked with gray. A severe and self-possessed air, and a large pale wart among the bushy hairs of one eyebrow? People talk about a young, pretty, beloved wife, dead in childbirth or some horrible accident, like a house fire or a railway disaster. There is no ground for this, but it adds interest. All he has told them is that his wife is dead.

He came to this part of the country looking for oil. The first oil well in the world was sunk in Lambton County, south of here, in the eighteen-fifties. Drilling for oil, Jarvis Poulter discovered salt.

He set to work to make the most of that. When he walks home from church with Almeda Roth he tells her about his salt wells. They are twelve hundred feet deep. Heated water is pumped down into them, and that dissolves the salt. The brine is pumped to the surface. It is poured into great evaporator pans over slow, steady fires, so that the water is steamed off and the pure, excellent salt remains. A commodity for which the demand will never fail.

"The salt of the earth," Almeda says.

"Yes," he says, frowning. He may think this disrespectful. She did not intend it so. He speaks of competitors in other towns who are following his lead and trying to hog the market. Fortunately, their wells are not drilled so deep, or their evaporating is not done so efficiently. There is salt everywhere under this land, but it is not so easy to come by as some people think.

Does that not mean, Almeda says, that there was once a great sea?

Very likely, Jarvis Poulter says. Very likely. He goes on to tell her about other enterprises of his — a brickyard, a lime kiln. and he explains to her how this operates, and where the good clay is found. He also owns two farms, whose woodlots supply the fuel for his operations.

And among the couples strolling home from church on a recent, sunny Sabbath morning we noted a certain salty gentleman and literary lady, not perhaps in their first youth but by no means blighted by the frosts of age. May we surmise?

This kind of thing pops up in the *Vidette* all the time.

May they surmise, and is this courting? Almeda Roth has a bit of money, which her father left her, and she has her house. She is not too old to have a couple of children. She is a good enough housekeeper, with the tendency toward fancy iced cakes and decorated tarts which is seen fairly often in old maids. (Honourable mention at the Fall Fair.) There is nothing wrong with her looks, and naturally she is in better shape than most married women of her age, not having been loaded down with work and children. But why was she passed over in her earlier, more marriageable years, in a place that needs women to be partnered and fruitful? She was a rather gloomy girl — that may have been the trouble. The deaths of her brother and sister and then of her mother, who lost her reason, in fact, a year before she died, and lay in her bed talking nonsense — those weighed on her, so she was not lively company. And all that reading and poetry — it seemed more of a drawback, a barrier, an obsession, in the young girl than in the middle-aged woman, who needed something, after all, to fill her time. Anyway, it's five years since her book was published, so perhaps she has got over that. Perhaps it was the proud, bookish father, encouraging her?

Everyone takes it for granted that Almeda Roth is thinking of Jarvis Poulter as a husband and would say yes if he asked her. And she is thinking of him. She doesn't want to get her hopes up too much, she doesn't want to make a fool of herself. She would like a signal. If he attended church on Sunday evenings, there would be a chance, during some months of the year, to walk home after dark. He would carry a lantern. (There is as yet no street lighting in town). He would swing the lantern to light the way in front of the lady's feet and observe their narrow and delicate shape. He might catch her arm as they step off the boardwalk. But he does not go to church at night.

Nor does he call for her, and walk with her *to* church on Sunday mornings. That would be a declaration. He walks her home, past his gate as far as hers; he lifts his hat then and leaves her. She does not invite him to come in — a woman living alone could never do such a thing. As soon as a man and woman of almost any age are alone together within four walls, it is assumed that anything may happen. Spontaneous combustion, instant fornication, an attack of passion. Brute instinct, triumph of the senses. What possibilities men and women must see in each other to infer such dangers. Or, believing in the dangers, how often they must think about the possibilities.

When they walk side by side she can smell his shaving soap, the barber's oil, his pipe tobacco, the wool and linen and leather smell of his manly clothes. The correct, orderly, heavy clothes are like those she used to brush and starch and iron for her father. She misses that job — her father's appreciation, his dark, kind authority. Jarvis Poulter's garments, his smell, his movement, all cause the skin on the

Alice

side of her body next to him to tingle hopefully, and a meek shiver raises the hairs on her arms. Is this to be taken as a sign of love? She thinks of him coming into her — their — bedroom in his long underwear and his hat. She knows this outfit is ridiculous, but in her mind he does not look so; he has the solemn effrontery of a figure in a dream. He comes into the room and lies down on the bed beside her, preparing to take her in his arms. Surely he removes his hat? She doesn't know, for at this point a fit of welcome and submission overtakes her, a buried gasp. He would be her husband.

One thing she has noticed about married women, and that is how many of them have to go about creating their husbands. They have to start ascribing preferences, opinions, dictatorial ways. Oh, yes, they say, my husband is very particular. He won't touch turnips. He won't eat fried meat. (Or he will only eat fried meat). He likes me to wear blue (brown) all the time. He can't stand organ music. He hates to see a woman go out bareheaded. He would kill me if I took one puff of tobacco. This way, bewildered, sidelong-looking men are made over, made into husbands, head of households. Almeda Roth cannot imagine herself doing that. She wants a man who doesn't have to be made, who is firm already and determined and mysterious to her. She does not look for companionship. Men — except for her father — seem to her deprived in some way, incurious. No doubt that is necessary, so that they will do what they have to do. Would she herself, knowing that there was salt in the earth, discover how to get it out and sell it? Not likely. She would be thinking about the ancient sea. That kind of speculation is what Jarvis Poulter has, quite properly, no time for.

Instead of calling for her and walking her to church, Jarvis Poulter might make another, more venturesome declaration. He could hire a horse and take her for a drive out to the country. If he did this, she would be both glad and sorry. Glad to be beside him, driven by him, receiving this attention from him in front of the world. And sorry to have the countryside removed for her — filmed over, in a way, by his talk and preoccupations. The countryside that she has written about in her poems actually takes diligence and determination to see. Some things must be disregarded. Manure piles, of course, and boggy fields full of high, charred stumps, and great heaps of brush waiting for a good day for burning. The meandering creeks have been straightened, turned into ditches with high, muddy banks. Some of the crop fields and pasture fields are fenced with big, clumsy uprooted stumps, others are held in a crude stitchery of rail fences. The trees have all been cleared back to the woodlots. And the woodlots are all second growth. No trees along the roads or lanes or around the farmhouses, except a few that are newly planted, young and weedy looking. Clusters of log barns — the grand barns that are to dominate the countryside for the next hundred years are just beginning to be built — and mean-looking log houses, and every four or five miles a ragged little shop. A raw countryside just wrenched from the forest, but swarming with people. Every hundred acres is a farm, every farm has a family, most families have ten or twelve children. (This is the country that will send out wave after wave of settlers — it's already starting to send them — to northern Ontario and the West). It's true that you can gather wildflowers in spring in the woodlots, but you'd have to walk through herds of horned cows to get to them.

IV
The Gypsies have departed.
Their camping-ground is bare.
Oh, boldly would I bargain now
At the Gypsy Fair.

Almeda suffers a good deal from sleeplessness, and the doctor has given her bromides and nerve medicine. She takes the bromides, but the drops gave her dreams that were too vivid and disturbing, so she has put the bottle by for an emergency. She told the doctor her eyeballs felt dry, like hot glass, and her joints ached. Don't read so much, he said, don't study; get yourself good and tired out with housework, take exercise. He believes that her troubles would clear up if she got married. He believes this in spite of the fact that most of his nerve medicine is prescribed for married women.

So Almeda cleans house and helps clean the church, she lends a hand to friends who are wallpapering or getting ready for a wedding, she bakes one of her famous cakes for the Sunday-school picnic. On a hot Saturday in August she decides to make some grape jelly. Little jars of grape jelly will make fine Christmas presents, or offerings to the sick. But she started late in the day and the jelly is not made by nightfall. In fact, the hot pulp has just been dumped into the cheesecloth bag, to strain out the juice. Almeda drinks some tea and eats a slice of cake with butter (a childish indulgence of hers), and that's all she wants for supper. She washes her hair at the sink and sponges off her body, to be clean for Sunday. She doesn't light a lamp. She lies down on the bed with the window wide open and a sheet just up to her waist, and she does feel wonderfully tired. She can even feel a little breeze.

When she wakes up, the night seems fiery hot and full of threats. She lies sweating on her bed, and she has the impression that the noises she hears are knives and saws and axes — all angry implements chopping and jabbing and boring within her head. But it isn't true. As she comes further awake she recognizes the sounds that she has heard sometimes before — the fracas of a summer Saturday night on Pearl Street. Usually the noise centres on a fight. People are drunk, there is a lot of protest and encouragement concerning the fight, somebody will scream "Murder!" Once, there was a murder. But it didn't happen in a fight. An old man was stabbed to death in his shack, perhaps for a few dollars he kept in the mattress.

She gets out of bed and goes to the window. The night sky is clear, with no moon and with bright stars. Pegasus hangs straight ahead, over the swamp. Her father taught her that constellation — automatically, she counts its stars. Now she can make out distinct voices, individual contributions to the row. Some people, like herself, having evidently been wakened from sleep. "Shut up!" they are yelling. "Shut up that caterwauling or I'm going to come down and tan the arse off yez!"

But nobody shuts up. It's as if there were a ball of fire rolling up Pearl Street, shooting off sparks — only the fire is noise, it's yells and laughter and shrieks and curses, and the sparks are voices that shoot off alone. Two voices gradually distinguish themselves — a rising and falling howling cry and a steady throbbing, low-pitched stream of abuse that contains all those words which Almeda associates with danger and depravity and foul smells and disgusting sights. Someone — the person crying out, "Kill me! Kill me now!" — is being beaten. A woman is being beaten. She keeps crying, "Kill me!

Kill me!" and sometimes her mouth seems choked with blood. Yet there is something taunting and triumphant about her cry. There is something theatrical about it. And the people around are calling out, "Stop it! Stop that!" or "Kill her! Kill her!" in a frenzy, as if at the theatre or a sporting match or a prize fight. Yes, thinks Almeda, she has noticed that before — it is always partly a charade with these people; there is a clumsy sort of parody, an exaggeration, a missed connection. As if anything they did — even a murder — might be something they didn't quite believe but were powerless to stop.

Now there is the sound of something thrown — a chair, a plank? — and of a woodpile or part of a fence giving way. A lot of newly surprised cries, the sound of running, people getting out of the way, and the commotion has come much closer. Almeda can see a figure in a light dress, bent over and running. That will be the woman. She has got hold of something like a stick of wood or a shingle, and she turns and flings it at the darker figure running after her.

"Ah, go get her!" the voices cry. "Go baste her one!"

Many fall back now; just the two figures come on and grapple, and break loose again, and finally fall down against Almeda's fence. The sound they make becomes very confused — gagging, vomiting, grunting, pounding. Then a long, vibrating, choking sound of pain and self-abasement, self-abandonment, which could come from either or both of them.

Almeda has backed away from the window and sat down on the bed. Is that the sound of murder she has heard? What is to be done, what is she to do? She must light a lantern, she must go downstairs and light a lantern — she must go out into the yard, she must go downstairs. Into the yard. The lantern. She falls over on her bed and pulls the pillow to her face. In a minute. The stairs, the lantern. She sees herself already down there, in the back hall, drawing the bolt of the back door. She falls asleep.

She wakes, startled, in the early light. She thinks there is a big crow sitting on her windowsill, talking in a disapproving but unsurprised way about the events of the night before. "Wake up and move the wheelbarrow!" it says to her, scolding, and she understands that it means something else by "wheelbarrow" — something foul and sorrowful. Then she is awake and sees that there is no such bird.

She gets up at once and looks out the window.

Down against her fence there is a pale lump pressed — a body.

Wheelbarrow.

She puts a wrapper over her nightdress and goes downstairs. The front rooms are still shadowy, the blinds down in the kitchen. Something goes *plop, plup,* in a leisurely, censorious way, reminding her of the conversation of the crow. It's just the grape juice, straining overnight. She pulls the bolt and goes out the back door. Spiders have draped their webs over the doorway in the night, and the hollyhocks are drooping, heavy with dew. By the fence, she parts the sticky hollyhocks and looks down and she can see.

A woman's body heaped up there, turned on her side with her face squashed down into the earth. Almeda can't see her face. But there is a bare breast let loose, brown nipple pulled long like a cow's teat, and a bare haunch and leg, the haunch bearing a bruise as big as a sunflower. The unbruised skin is greyish, like a plucked, raw drumstick. Some kind of nightgown or all-purpose dress she has on. Smelling of vomit. Urine, drink, vomit.

Barefoot, in her nightgown and flimsy wrapper, Almeda runs away. She runs around the side of her house between the apple trees and the veranda; she opens the front gate and flees down Dufferin Street to Jarvis Poulter's house, which is the nearest to hers. She slaps the flat of her hand many times against the door.

"There is the body of a woman," she says when Jarvis Poulter appears at last. He is in his dark trousers, held up with braces, and his shirt is half unbuttoned, his face unshaven, his hair standing up on his head. "Mr. Poulter, excuse me. A body of a woman. At my back gate."

He looks at her fiercely. "Is she dead?"

His breath is dank, his face creased, his eyes bloodshot.

"Yes. I think murdered," says Almeda. She can see a little of his cheerless front hall. His hat on a chair. "In the night I woke up. I heard a racket down on Pearl Street," she says, struggling to keep her voice low and sensible. "I could hear this — pair. I could hear a man and a woman fighting."

He picks up his hat and puts it on his head. He closes and locks the front door, and puts the key in his pocket. They walk along the boardwalk and she sees that she is in her bare feet. She holds back what she feels a need to say next — that she is responsible, she could have run out with a lantern, she could have screamed (but who needed more screams?), she could have beat the man off. She could have run for help then, not now.

They turn down Pearl Street, instead of entering the Roth yard. Of course, the body is still there. Hunched up, half bare, the same as before.

Jarvis Poulter doesn't hurry or halt. He walks straight over to the body and looks down at it, nudges the leg with the toe of his boot, just as you'd nudge a dog or a sow.

"You," he says, not too loudly but firmly, and nudges again.

Almeda tastes bile at the back of her throat.

"Alive," says Jarvis Poulter, and the woman confirms this. She stirs, she grunts weakly.

Almeda says, "I will get the doctor." If she had touched the woman, if she had forced herself to touch her, she would not have made such a mistake.

"Wait," says Jarvis Poulter. "Wait. Let's see if she can get up."

"Get up, now," he says to the woman. "Come on. Up, now. Up."

Now a startling thing happens. The body heaves itself onto all fours, the head is lifted — the hair all matted with blood and vomit — and the woman begins to bang this head, hard and rhythmically, against Almeda Roth's picket fence. As she bangs her head she finds her voice, and belts out an open-mouthed yowl, full of strength and what sounds like an anguished pleasure.

"Far from dead," says Jarvis Poulter. "And I wouldn't bother the doctor."

"There's blood," says Almeda as the woman turns her smeared face.

"From her nose," he says. "Not fresh." He bends down and catches the horrid hair close to the scalp to stop the head banging.

"You stop that now," he says. "Stop it. Gwan home now. Gwan home, where you belong." The sound coming out of the woman's mouth has stopped. He shakes her head slightly, warning her, before he lets go of her hair. "Gwan home!"

Released, the woman lunges forward, pulls herself to her feet. She can walk. She weaves and stumbles down the street, making intermittent, cautious noises of protest. Jarvis Poulter watches her for a moment to make sure that she's on her way. Then he finds a large burdock leaf, on which he wipes his hand. He says, "There goes your dead body!"

The back gate being locked, they walk around to the front. The front gate stands open. Almeda still feels sick. Her abdomen is bloated; she is hot and dizzy.

"The front door is locked," she says faintly, "I came out by the kitchen." If only he would leave her, she could go straight into the privy. But he follows her. He follows her as far as the back door and into the back hall. He speaks to her in a tone of harsh joviality that she has never before heard from him. "No need for alarm," he says. "It's only the consequences of drink. A lady oughn't to be living alone so close to a bad neighbourhood." He takes hold of her arm just above the elbow. She can't open her mouth to speak to him, to say thank you. If she opened her mouth she would retch.

What Jarvis Poulter feels for Almeda Roth at this moment is just what he has not felt during all those circumspect walks and all his own solitary calculations of her probable worth, undoubted respectability, adequate comelinesss. He has not been able to imagine her as a wife. Now that is possible. He is sufficiently stirred by her loosened hair — prematurely gray but thick and soft — her flushed face, her light clothing, which nobody but a husband should see. And by her indiscretion, her agitation, her foolishness, her need?

"I will call on you later," he says to her. "I will walk with you to church."

At the corner of Pearl and Dufferin streets last Sunday morning there was discovered, by a lady resident there, the body of a certain woman of Pearl Street, thought to be dead but only, as it turned out, dead drunk. She was roused from her heavenly — or otherwise — stupor by the firm persuasion of Mr. Poulter, a neighbour and a Civil Magistrate, who had been summoned by the lady resident. Incidents of this sort, unseemly, troublesome, and disgraceful to our town, have of late become all too common.

V

I sit at the bottom of sleep,
As on the floor of the sea.
And fanciful Citizens of the Deep
Are graciously greeting me.

As soon as Jarvis Poulter has gone and she has heard her front gate close, Almeda rushes to the privy. Her relief is not complete, however, and she realizes that the pain and fullness in her lower body come from an accumulation of menstrual blood that has not yet started to flow. She closes and locks the back door. Then, remembering Jarvis Poulter's words about church, she writes on a piece of paper, "I am not well, and wish to rest today." She sticks this firmly into the outside frame of the little window in the front door. She locks that door, too. She is trembling, as if from a great shock or danger. But she builds a fire, so that she can make tea. She boils water, measures the tea leaves, makes a large pot of tea, whose steam and smell sicken her further. She pours out a cup while the tea is still quite weak and adds to it several dark drops of nerve medicine. She sits to drink it without raising the kitchen blind. There, in the middle of the floor, is the cheesecloth bag hanging on its broom handle between the two chair backs. The grape pulp and juice has stained the swollen cloth a dark purple. *Plop, plup* into the basin beneath. She can't sit and look at such a thing. She takes her cup, the teapot, and the bottle of medicine into the dining room.

   She is still sitting there when the horses start to go by on the way to church, stirring up clouds of dust. The roads will be getting hot as ashes. She is there when the gate is opened and a man's confident steps sound on her veranda. Her hearing is so sharp she seems to hear the paper taken out of the frame and unfolded — she can almost hear him reading it, hear the words in his mind. Then the footsteps go the other way, down the steps. The gate closes. An image comes to her of tombstones — it makes her laugh. Tombstones are marching down the street on their little booted feet, their long bodies inclined forward, their expressions preoccupied and severe. The church bells are ringing.

   Then the clock in the hall strikes twelve and an hour has passed.

The house is getting hot. She drinks more tea and adds more medicine. She knows that the medicine is affecting her. It is responsible for her extraordinary langour, her perfect immobility, her unresisting surrender to her surrounding. That is all right. It seems necessary.

Her surroundings — some of her surroundings — in the dining room are these: walls covered with dark green garlanded wallpaper, lace curtains and mulberry velvet curtains on the windows, a table with a crocheted cloth and a bowl of wax fruit, a pinkish-grey carpet with nosegays of blue and pink roses, a sideboard spread with embroidered runners and holding various patterned plates and jugs and the silver tea things. A lot of things to watch. For everyone of these patterns, decorations, seems charged with life, ready to move and flow and alter. Or possibly to explode. Almeda Roth's occupation throughout the day is to keep an eye on them. Not to prevent their alteration so much as to catch them at it — to understand it, to be a part of it. So much is going on in this room that there is no need to leave it. There is not even the thought of leaving it.

Of course, Almeda in her observations cannot escape words. She may think she can, but she can't. Soon this glowing and swelling begins to suggest words — not specific words but a flow of words somewhere, just about ready to make themselves known to her. Poems, even. Yes, again, poems. Or one poem. Isn't that the idea — one very great poem that will contain everything and, oh, that will make all the other poems, the poems she has written, inconsequential, mere trial and error, mere rags? Stars and flowers and birds and trees and angels in the snow and dead children at twilight — that is not the half of it. You have to get in the obscene racket on Pearl Street and the polished toe of Jarvis Poulter's boot and the plucked-chicken haunch with its blue-black flower. Almeda is a long way now from human sympathies or fears or cozy household considerations. She doesn't think about what could be done for that woman or about keeping Jarvis Poulter's dinner warm and hanging his long underwear on the line. The basin of grape juice has overflowed and is running over her kitchen floor, staining the boards of the floor, and the stain will never come out.

She has to think of so many things at once — Champlain and

the naked Indians and the salt deep in the earth but as well as the salt the money, the money-making intent brewing forever in heads like Jarvis Poulter's. Also, the brutal storms of winter and the clumsy and benighted deeds on Pearl Street. The changes of climate are often violent, and if you think about it there is no peace even in the stars. All this can be borne only if it is channelled into a poem, and the word "channelled" is appropriate, because the name of the poem will be — it is — "The Meneseteung." The name of the poem is the name of the river. No, in fact it is the river, the Meneseteung, that is the poem — with its deep holes and rapids and blissful pools under the summer trees and its grinding blocks of ice thrown up at the end of winter and its desolating spring floods. Almeda looks deep, deep into the river of her mind and into the tablecloth, and she sees the crocheted roses floating. They look bunchy and foolish, her mother's crocheted roses — they don't look much like real flowers. But their effort, their floating independence, their pleasure in their silly selves, does seem to her so admirable. A hopeful sign. Meneseteung.

She doesn't leave the room until dusk, when she goes out to the privy again and discovers that she is bleeding, her flow has started. She will have to get a towel, strap it on, bandage herself up. Never before, in health, has she passed a whole day in her nightdress. She doesn't feel any particular anxiety about this. On her way through the kitchen she walks through the pool of grape juice. She knows that she will have to mop it up, but not yet, and she walks upstairs leaving purple footprints and smelling her escaping blood and the sweat of her body that has sat all day in the closed hot room.

No need for alarm.

For she hasn't thought that crocheted roses could float away or that tombstones could hurry down the street. She doesn't mistake that for reality, and neither does she mistake anything else for reality, and that is how she knows that she is sane.

VI

I dream of you by night,
I visit you by day.
Father, Mother,
Sister, Brother,
Have you no word to say?

April 22, 1903. At her residence, on Tuesday last, between three and four o'clock in the afternoon there passed away a lady of talent and refinement whose pen, in days gone by, enriched our local literature with a volume of sensitive, eloquent verse. It is a sad misfortune that in later years the mind of this fine person had become somewhat clouded and her behaviour, in consequence, somewhat rash and unusual. Her attention to decorum and to the care and adornment of her person had suffered, to the degree that she had become, in the eyes of those unmindful of her former pride and daintiness, a familiar eccentric, or even, sadly, a figure of fun. But now all such lapses pass from memory and what is recalled is her excellent published verse, her labours in former days in the Sunday school, her dutiful care of her parents, her noble womanly nature, charitable concerns, and unfailing religious faith. Her last illness was of mercifully short duration. She caught cold, after having become thoroughly wet from a ramble in the Pearl Street bog. (It has been said that some urchins chased her into the water, and such is the boldness and cruelty of some of our youth, and their observed persecution of this lady, that the tale cannot be entirely discounted.) The cold developed into pneumonia and she died, attended at the last by a former neighbour, Mrs. Bert (Annie) Friels, who witnessed her calm and faithful end.

January, 1904. One of the founders of our community, an early maker and shaker of this town, was abruptly removed from our midst on Monday morning last, whilst attending to his correspondence in the office of his company. Mr. Jarvis Poulter possessed a keen and lively commercial spirit, which was instrumental in the creation of not one but several local enterprises, bringing the benefits of industry, productivity, and employment to our town.

I looked for Almeda Roth in the graveyard. I found the family stone. There was just one name on it — Roth. Then I noticed two flat stones in the ground, a distance of a few feet — six feet? — from the upright stone. One of these said "Papa," the other "Mama." Farther out from these I found two other flat stones, with the names William and Catherine on them. I had to clear away some overgrowing grass and dirt to see the full name of Catherine. No birth or death dates for anybody, nothing about being dearly beloved. It was a private sort of memorializing, not for the world. There were no roses, either

— no sign of a rosebush. But perhaps it was taken out. The groundskeeper doesn't like such things, they are a nuisance to the lawnmower, and if there is nobody left to object he will pull them out.

I thought that Almeda must have been buried somewhere else. When this plot was bought — at the time of the two children's death — she would still have been expected to marry, and to lie finally beside her husband. They might not have left room for her. Then I saw that the stones in the ground fanned out from the upright stone. First the two for the parents, then the two for the children, but these were placed in such a way that there was room for a third, to complete the fan. I paced out from "Catherine" the same number of steps that it took to get from "Catherine" to "William," and at this spot I began pulling grass and scrabbling in the dirt with my bare hands. Soon I felt the stone and knew that I was right. I worked away and got the whole stone clear and I read the name "Meda." There it was with the others, staring at the sky.

I made sure I had got to the edge of the stone. That was all the name there was — Meda. So it was true that she was called by that name in the family. Not just in the poem. Or perhaps she chose her name from the poem, to be written on her stone.

I thought that there wasn't anybody alive in the world but me who would know this, who would make the connection. And I would be the last person to do so. But perhaps this isn't so. People are curious. A few people are. They will be driven to find things out, even trivial things. They will put things together, knowing all along that they may be mistaken. You see them going around with notebooks, scraping the dirt off gravestones, reading microfilm, just in the hope of seeing this trickle in time, making a connection, rescuing one thing from the rubbish.

Worlds that

FLOURISH

BEN OKRI

Ben Okri was born in Nigeria in 1959. He won the Booker Prize in 1991. He has redefined realism in his writing as: "all that's there — what we see and what we don't see — I don't think I've moved away from realism. I think I've just moved deeper into it." Of the urgency in his writing, he analogizes, "In the martial arts you pay for every mistake you make. You pay for loss of attention, you pay for taking anything for granted. In a piece of writing, it should be the same way ..."

I WAS AT WORK ONE DAY WHEN A MAN CAME up to me and asked me my name. For some reason I couldn't tell it to him immediately and he didn't wait for me to get around to it before he turned and walked away. At lunch-time I went to the bukka to eat. When I got back to my desk someone came and told me that half the workers in the department had been sacked. I was one of them.

I had not been working long in the department and I left the job without bitterness. I packed my things that day and sorted out the money that was owed me. I got into my battered little car and drove home. When I arrived I parked my car three streets from where I lived, because the roads were bad. As I walked home the sight of tenements and zinc huts made me dizzy. Swirls of dust came at me from the untarred roads. Everything shimmered like mirages in an omnipotent heat.

Later in the evening I went out to buy some cooked food. On my way back a neighbour came to me and said: "How are you?"

"Fine," I said.

"Are you sure you are fine?"

"Yes. Why do you ask?"

"Well," said the neighbour, "it's because you go around as if you don't have any eyes."

"What do you mean?"

"Since your wife died you've stopped using your eyes. Haven't you noticed that most of the compound people are gone?"

"Gone where?"

"Run away. To safety."

"Why haven't you gone?"

"I'm happy here."

"So am I," I said, smiling. I went to my room.

Barely two hours after the conversation with my neighbour there was a knock on my door. I opened it and three men pushed their way in. Two of them carried machetes and the third had a gun. They weren't nasty or brutal. They merely asked me to sit quietly on the bed and invited me to watch them if I wanted. I watched them as they cleaned my room of my important possessions and took what money they could find. They chatted to me about how bad the roads were and how terrible the government was and how there were so many checkpoints around. While they chatted they bundled my things into a heap; and carried them out to their lorry as though they were merely helping me to move. When they finished the man with the gun said:

"This is what we call scientific robbery. If you so much as cough after we've gone I will shoot out your eyes, you hear?"

I nodded. He left with a smile. A moment later I heard their lorry driving off down the untarred road. I rushed out and they were gone. I came back to my room to decide what next to do. I couldn't inform the police immediately because the nearest station was miles away and even if I did I couldn't really expect them to do anything. I sat on the bed and tried to convince myself that I was quite fortunate to still have the car and some money in the bank. But as it turned out I wasn't even allowed to feel fortunate. Not long after the thieves had left there was another knock on my door. I got up to open it when five soldiers with machine-guns stepped into the room. Apparently the thieves had been unable to get away. They were stopped at a checkpoint and to save their own necks they told the soldiers that I was their accomplice. Without ceremony, and with a great deal of roughness, the soldiers dragged me to their jeep. Visions of being executed as an armed robber at the beach filled me with vertigo. I told the soldiers that I was the one who was robbed but the soldiers began to beat me because it seemed to them I was trying to insult their intelligence with such a transparent lie. As they took me away, with their guns prodding my back, my neighbour came out of his room. When he saw the soldiers with me he said:

"I told you that you don't have eyes."

Then he went to one of the soldiers and, to my astonishment, said:

"Mr. Soldier, I hope you treat him as he deserves. I always thought something was wrong with his head."

The soldiers took us to the nearest police station and we were all locked in the same cell. The real thieves, who seemed to find it all amusing, kept smiling at me. At night the soldiers came and beat us up with whips when we refused to confess anything. Then in the morning some policemen took us outside and made us strip naked and commanded us to face the street. The people that went past looked at us and hurried on. I shouted of my innocence and the policemen told me to shut up. We stayed out facing the whole world in our nakedness for most of the day. The children laughed at us. The women studied us. Photographers came and flashed their cameras in our eyes. When night fell a policeman came and offered me the opportunity to bribe my way out of trouble. I burned all over and my eyes were clogged with dust. I told him I had to go to the bank first. The thieves paid their dues and were freed. I stayed in a cell crammed with men screaming all night. In the morning one of the soldiers accompanied me to the bank. I drew out some money and paid my dues. I went home and slept for the rest of that day.

In the morning I went to have a shower. Going through the compound I was struck by the absence of communal noises. No music came from the rooms. No children cried. There were no married couples arguing and shouting behind red curtains. There were chickens and rats in the backyard. My neighbour came out of the toilet and smiled when he saw me.

"So they have released you," he said, regretfully.

"You are a wicked man," I shouted.

"People don't go out anymore," he said, coolly ignoring me. "It's very quiet. I like it this way."

"Why were you so wicked to me?"

"I don't trust people who don't have eyes."

"I might have been executed."

"Are you better than those who have been?"

I stared at him in disbelief. He went and washed his hands at the pump and dried them against his trousers. He pushed past me and went to his room. A moment later I saw him going out.

I still felt sleepy even after my shower. I went to my room and got dressed. Then I went to the front of the compound. I sat on a bench and looked at the street. The churches around were not having their usual prayers and songs over loudspeakers. The muezzin was silent. The street was deserted. There were no signs of panic. The stalls still had their display of goods and the shops were open, but there was no one around. There were a lot of birds in the air, circling the aerials. Somewhere in the distance a radio had been left on. Across the street a goat wandered around the roots of a tree. The cocks didn't crow. After a while all I heard inside me was a confused droning, my incomprehension. Something had been creeping on us all along and now that the street was empty I couldn't even see what it was. I sat outside, fighting the mosquitoes, till it became dark. Then it dawned on me that something had happened to time. I seemed to be sitting in an empty space without history. The wind wasn't cooling. And then suddenly all the lights went out. It was as if the spirit of the world had finally died. The black-out lasted a long time.

For many days I wandered about in the darkness of the city. I drove around in the day looking for jobs. Everywhere I went workers were being sacked in great numbers. There were no strikes. Sometimes I listened to the Head of State's broadcasts on the radio. He spoke about austerity, about tightening the national belt, and about a great future. He sounded very lonely, as though he were talking in a vast and empty room. After his broadcasts music was played. The music sounded also as if it were played in an empty space.

In the evenings I went around looking for friends. They had gone and no explanations or forwarding addresses were given. When I went to their compounds I was surprised at how things had changed. The decay of the compounds seemed to have accelerated. Doors were left open. Cobwebs hung over the compound fronts. Outside the house of a friend I saw a boy staring at me with frightened eyes. When I started to ask him of the whereabouts of my friend, he got up and ran. I went back to my car and drove around the city, looking for people that I knew. Then I really began to notice things. There were people scattered in places of the city. There seemed no panic on their faces. It began to occur to me that the world was emptying out. When I took a closer look at the people a strange thought came to me: they seemed like sleep-walkers. I stopped the car and went amongst them to get a closer look, to talk with them, and find out exactly what was happening. (The radios and newspapers had long stopped giving information). I went out into the street and approached a woman who was frying yams at the roadside. She looked at me with burning, suspicious eyes.

"What is happening to the country?" I asked her.

"Nothing is happening."

"Where has everyone gone?"

"No one has gone anywhere. Why are you asking me? Go and ask someone else."

As I turned to go the fire flared up, illuminating her face. And on her face I saw a sloping handwriting. On her forehead and on her cheeks there were words. Then I noticed that her hands were also covered in handwriting. I drew closer to read the words, but she began screaming. I heard the ironclad boots of soldiers running down the streets towards us. I hurried to my car and drove off.

As I went home I noticed that a lot of the people in the streets had handwriting on their faces. I couldn't understand why I hadn't noticed it before. And then I was suddenly overcome with the notion that my neighbour had words on his face. I drove home hurriedly.

It was dark by the time I arrived. I couldn't risk having the car three streets away, so I parked it outside the compound. I think it was with that act of caution that the thought of fleeing first occurred to me. The birds had increased over our street. The radio was still on somewhere in the distance. Its battery was getting weaker. The wind whistled though the compounds. Stray dogs roamed down the street. I sat outside and waited for my neighbour. When he didn't come back for a long time I went and knocked on his door. There was no reply. I went to my room and ate, and then I went and sat outside again. I listened to the radio dying. I listened to the thin military voices. The night got darker and still my neighbour didn't return. I listened to the wind straining the branches of the trees. Stray cats eyed me in the dark. I went to my room and I slept that night with the feeling that something was breaking on my consciousness. When I woke up in the morning I noticed that the Head of State's lonely face kept slipping into my mind. I had a shower and ate and went and knocked on my neighbour's door. He still hadn't got back.

I prepared to go out but thunder sounded in the sky. By the afternoon it had started to rain. The street swelled with water. The gutters overran. The rain poured into the open doors of the rooms and fell on the stalls with their undisturbed display of goods and beat down on the clothes that had been left hanging. The wind blew very hard and shook our roof. The branches of a tree strained and then cracked. From afar I could see smoke above the houses. The rain poured down unceasingly for two days. My neighbour still didn't return. The water went up to the bumper of my car. The rain finally extinguished the distant radio. The Head of State made desperate broadcasts about cleaning the national stables. I sat in my room, imprisoned by the rain. I listened to the water endlessly falling. My roof began to leak. I heard a cat wailing above the steady din. Sometimes the rain accelerated in its fall and managed to obliterate both time and memory. It soon seemed as if it had always been raining. With the city empty of people, I began to hear broadcasts in the rain. And then in the evening of the second day, a realization came upon me. I went to the window, my ears reverberating with persistently dripping water, and looked out. That was when I discovered I had temporarily lost the names of things.

I stayed indoors till the rain stopped. Then I stayed in another day, to enable the water to sink into the swollen earth. I went and tried my neighbour's door several times and then I went into his room. Nothing had been disturbed, but he seemed to have altogether vanished. On the fourth day I ventured down our street and witnessed the proliferation of disasters. Trees had fallen. Houses had crumbled before the force of the wind and rain. Dead cats floated in the gutters. There were no birds in the air. I went back to my room. My head jostled with signs. I got out my box and stuffed it full of my papers and clothes. I packed all my food into the back of the car. I left my door open. I tried my neighbour's room for the final time. I got into my car and set out on a journey without a destination through the vast, uncultivated country.

It wasn't easy going out of the city. There were so many roadblocks and soldiers were all over the place. They stopped me and searched the car. At every one of the roadblocks the soldiers commented on the food I had at the back. They asked where I was going. I told them I was going to visit my mother who was ill in the village. Then they would ask if I thought that people were hungry. When I said no, the soldiers would take some of my food and wave me on. By the time I cleared through the last roadblock I had very little food left. But that wasn't what worried me. What made me anxious, as I drove through the forests, was that the car kept giving me trouble. It would stall and I had to sit at the wheel and wait for the engine to cool. When it did start, and move, it did so erratically. The car would suddenly, it seemed, start driving me. It picked up speed, and slowed down, of its own inscrutable volition.

I drove for a long time down the winding forest road. I managed to cross a wooden bridge that had been partly devastated by rain. For long periods of time I heard only the purring sound of the car. Sometimes it seemed as if I were driving on one spot. The road and forests didn't seem to change. I crossed the same partly devastated bridge several times. I got tired of driving without seeming to be moving. I stopped and locked all the doors and got some sleep.

I felt better when I woke up. I was driving for a while when I felt that I had broken the sameness of the journey. Mountain ranges, plateaus of ambergris rocks, and precipices, appeared all around me. I passed a clay-coloured anthill. I slowed down for a pack of hyenas to cross the road. I came to a petrol shack. The door was open. There were dirty barrels of petrol and diesel oil in the front yard. I stopped the car and parked. I passed the greasy hand-pump and knocked on the door. An old man came out. He had a pair of grey braces over a black shirt and he wore filthy khaki trousers. He was barefoot.

"You're the first person I've seen for a long time," he said.

I asked him to fill the tank. He didn't say anything to me as he did so. I changed the water in the radiator. He didn't have any brake fluid. I sat on a bench and listened to the insects of the forest while he slowly and painstakingly looked the car over and tuned the engine.

"How do you manage to live here?"

"I manage. I like it."

I paid him. As I was getting into the car the old man said:

"Don't go that way. I haven't seen any vehicles coming back. Stay where you can be happy."

I nodded, smiling. I shut the door and started the car. As I moved away I waved at him. He didn't wave back. He stared at the car motionlessly. I drove on into the forest.

Further along I ran over a goat that had been crossing the road. I felt the wheels bump over its body and I stopped. The goat jerked on the tarmac. When I came out of the car I heard violent noises and saw people emerging from the forest and rushing towards me. The men had machetes and the women held long pieces of firewood. I ran back to the car, but when I started it the engine only whined. The people pounced on the car and smashed the bodywork with their machetes and firewood. They broke the windows and several hands reached for my face. The car started, suddenly, and I sped off with a few hands still grasping for my eyes. I swerved both ways and people fell off and I drove on without looking back. Afterwards I saw blood and bits of flesh on the jagged, broken windows.

And then it was as if the rain that had fallen in the city began to catch up with me, intensified. The forest reverberated with thunder. Lightning struck in the trees. The leaves were blown into frenzies by the relentless wind. The car kept swerving and sometimes it was as if the wind was blowing the car on, lifting it at the back. Sometimes I did not feel that the wheels were on the road. I drove on air. I drove on through the torrential rain. There were trees swaying and leaves flapping everywhere. And then there was water pouring on the trees everywhere. Now and again someone would emerge, soaking, from the forest and would run across the road and wave for me to stop. I did not stop for anybody, or for any reason. I drove on in demented concentration. Soon my eyes got tired. I was thrashed by the rain and all I could see was the windscreen and the forests distorted in the rain. I found it difficult to blink and when I did I felt the blankness pulling me into sleep. I would wake up to find myself veering off the road. I managed to sleep while driving.

When night came thickly over the forest I couldn't separate the darkness from the rain. Occasionally I saw a flash behind me which I thought belonged to a car. I adjusted the mirror and in the crack of a second I saw my face. Thunder broke and exploded in front of me. A moment later there was a forked, incandescent flash which lit up the handwriting on my face. I negotiated a bend and heard a deafening crash in the forest. Something shattered my windscreen and I drove wide-eyed into the darkness. Insects flew into my face. Wind, rain, and bits of glass momentarily blinded me. Then I saw that a tree had fallen across the road ahead of me. The car spun into the vortex of leaves and branches. And then there was stillness. For a long moment it was completely dark. I couldn't hear, see, or feel anything. And then I heard the whirring engine and the insistent din of insects and rain.

I tried to move, but I couldn't: I felt I had become entangled in the car. I heard magnified grating noises. I was covered in crumbly earth which seemed alive and which stung me. Something settled inside me and I extricated myself from the front seat effortlessly. When I was out of the wreckage I saw that the car had run into a large anthill. There were ants everywhere. I pushed on through the rain. I couldn't find the road. I went on into the forest. I passed rocks flowering with lichen. I moved under the endless lattice of branches. Thorns of the forest cut into me. I didn't bleed.

I came to a river. When I swam across I noticed it was flowing in a direction opposite to how it seemed. As I came out on the other bank the water dried instantly on me. I went on through the undergrowth till I came to a village. At the entrance there were two palm trees growing upside down. I went between the trees and saw a man sitting on a chair outside a hut. When the man saw me his face lit up. He ululated suddenly and talking drums sounded at distances in the village. The man got up and rushed to me and embraced me:

"We've been waiting for you," he said.

"What do you mean?"

"We've been waiting for you."

"That can't be true."

The man looked quite offended at my remark, but he said:

"I have been sitting outside this hut for three months.

Waiting for you. I'm happy that you've made it. Come, the people of the village are expecting you."

He led the way.

"Why?"

"You'll find out."

I followed him silently. As we went on into the village, I noticed that there was a woman following us. Whenever I looked back she hid behind the trees and bushes.

"We've been cleaning up the village for your arrival," the man said.

We passed a skyscraper that reflected the sunlight like blinding glass sheets.

"That's where the meeting will take place."

The huts looked solid and clean with their white ochred walls. The iroko and baobab trees were neatly spaced. The bushes were lush. The air was scented with flamingo flowers.

We arrived at the village square when it occurred to me that the place was vaguely familiar. It was a very orderly and clean place. And then suddenly I realized that I couldn't see. I didn't hear the man leading me anymore. I heard singing and dancing all around. I panicked and started shouting. The dancing and singing stopped. I stood for a long time, casting about in the menacing silence. After a while, when I quietened down, I heard light footsteps coming towards me.

"Help me," I said.

Then a woman, who smelt of cloves, in a sweet voice, said:

"Be quiet and follow me."

I followed her till we came to a place that smelt of bark. She opened a door and we went in. She pulled up a stool for me. I could have been sitting on solid air for all I knew, but the woman's presence reassured me. I heard her moving about the place. She set down food for me. I ate. She set down drinks and I drank. Then she said:

"This will be your new home."

Then I heard the door shut. I soon fell asleep.

When I woke up I felt things coming out of my ears. Things were crawling all over me. I stood up and called out. The door opened and the woman came in and led me to the place where I had a wash.

After I had eaten, she sat near me and said:

"We heard you were coming. It took a long time."

"How did you hear?"

"You will find out."

"Why have you all been waiting for me?"

She was silent. Then she laughed and said:

"Didn't you know we have been waiting for you?"

"No."

"Didn't you know you were coming here?"

"No. But why?"

"To take your place in the assembly."

"What assembly?"

"We kept postponing the meeting because you hadn't arrived."

I grew weary of asking questions

"The people of the village have been anxious," she said.

"When is this meeting taking place?"

"Two days' time."

"Why not today?"

"The elders thought you needed time to rest and get used to the village. It's an important meeting."

"What is the meeting about?"

"You are tired. Get some sleep. If you need me call."

Then I heard the door open and shut again.

In the village everything had a voice and everything spoke at me. Sounds and voices assaulted me and my ears began to ache. Then slowly my sight returned. At first it was like seeing through milk. When my vision cleared, the voices stopped. Then I saw the village as I had not seen it before.

I went out of the place I was staying and walked around in bewilderment. Some of the people of the village had their feet facing backwards. I was amazed that they could walk. Some people came out of tree trunks. Some had wings, but they couldn't fly. After a while I got used to the strangeness of the people. I ceased to really notice their three legs and elongated necks. What I couldn't get used to were the huts and houses that were walled round with mirrors on the outside. I didn't see myself reflected in them as I went past.

Some people walked into the mirrors and disappeared. I couldn't walk into them.

After some time of moving around, I couldn't find my way back to where I stayed. I went about the village listening for the voice of the woman who had been taking care of me. I stopped at a communal water-pump and a woman came up to me and said:

"What are you doing here?"

"I'm lost."

"I'll take you back."

I followed her.

"So you can see now?" she asked, turning her head right round to me as she walked.

"Yes."

And then I had the distinct and absurd feeling that I knew her. She was a robust figure, with a face of jagged and familiar beauty. She wore a single flowerprint wrapper and was barefooted. Her skin was covered in native chalk. Her eyes radiated a strange light which dazzled like a green mirror.

"Who are you?"

She didn't answer my question. When we got to an obeche tree she opened a door on the trunk. Inside I saw a perfect interior, neat and compact and warm.

"I'm not going in there." I said.

She turned her head towards me, her face was expressionless.

"But this has been your new home,"she said.

"It can't be. It's too small."

She laughed almost affectionately.

"When you come in you will find it is large enough."

It was very spacious when I went in. I sat down on the wooden bed. She served me food in a half calabash. The rice seemed to move on the plate like several white maggots. I could have sworn it was covered in spider's webs. But it tasted sweet and was satisfying. The cup from which I was supposed to drink bled on the outside. After she had cleared the food from the table, I pretended to be asleep. Before she left I heard her say:

"Sleep well and regain your strength. The meeting is taking place tonight."

I sat up.

"Who are you?" I asked

She shut the door gently behind her.

I waited for some time before I got up and left the tree. I was intent on fleeing, but I didn't want to betray it. As I wandered round the village looking for the way out, I heard people dancing, I heard some disputing the village principles, I heard others reciting a long list of names, and I heard beautiful voices telling stories behind the trees. But I could not see any of the people.

And then as I passed a hut, from which came the high-pitched laughter of shy young girls, I noticed that a one-eyed goat was staring at me intently. I hurried on. Dogs and chickens gazed at me. I experienced the weird sensation that people were staring at me through the eyes of the animals. I passed the village shrine. In front of it there was the mighty statue of a god with big holes for eyes. I was convinced the god was spying on me.

I wandered for a long time looking for the exit. I heard disembodied voices saying that the big meeting would soon begin. The lights hadn't changed. I came to a frangipani tree full of white birds. Beyond the tree was the village square and beyond the square was the entrance. I pushed on till I came to the hut. Sitting on the chair outside the hut was a man who had three eyes on his face. He kept staring at me and I was forced to greet him.

"Don't greet me," he said.

He went on staring at me, as though he expected me to recognize him. His three eyes puzzled and disorientated me. But when I concentrated on the two normal eyes I suddenly did recognize him. He was my vanished neighbour.

"What are you doing here?" I asked.

"What do you think?"

"I don't know."

"A soldier shot me."

"Shot you?" I asked, surprised.

"Yes."

"Why?"

"To kill me. What are you doing here?"

"Me?"

"Yes."

"I don't know."

He laughed.

"They will tell you at the meeting."

"What is the meeting about?"

"Life and death."

"What life, what death?"

He laughed again, but more explosively. There was something about his mouth, the way his eyes moved, that gradually made things clear to me. I backed away in terror.

"You better not try and escape," he said maliciously.

That was all I needed. I ran towards the entrance and things got scrambled up as I ran. And then I found that I was moving not forwards, but backwards. I passed the white ochred huts and the blinding skyscraper. I heard the high-pitched scream of a woman. Talking drums sounded in frenzies. When I stopped and ran backwards, I found I was actually running forwards. Then I saw the woman who had screamed, and for the first time I recognized her as my dead wife. She tore after me in great distress. Men and women and disembodied voices came after me with their wings that didn't help them fly and their feet which were turned backwards. I fled past the trees that were upside down and the cornfields outside the village entrance. The cornplumes were golden and beautiful. The people of the village pursued me all the way to the boundary.

I crossed the river. Birds came at me from the forest. I ran for a long time without stopping till I came to my car that had smashed through the branches of the tree and devastated the anthill. I am not sure what happened next but when I came to I found myself in the wreckage of the car. I was covered with ants and they bit me mercilessly. The twisted wreck of metal seemed to have grown on me and I could feel my blood drying on the seat. There were cuts and broken glass on my face. I spent a very long time struggling to get out of the car. When I did I felt about as wrecked as the car and my body felt like it had already died. I staggered through the forest. I ate lemon grass leaves. As I pushed my way through the forest I became aware that I could see spirits. It was morning before I could find the main road. After a while of stumbling down the road I saw a car coming towards me. I stuck out my hand and waved furiously and was surprised when the car stopped. There was a young man at the wheel. He wound down his side window and I said:

"Don't go that way. Find where you can be happy."

But the young man looked me over, nodded, and drove straight on. Then I trudged on with the hope of reaching the old man's shack before I died.

THE SHAWL

CYNTHIA OZICK

STELLA, COLD, COLD, THE COLDNESS OF HELL. How they walked on the roads together, Rosa with Magda curled up between sore breasts, Magda wound up in the shawl. Sometimes Stella carried Magda. But she was jealous of Magda. A thin girl of fourteen, too small, with thin breasts of her own, Stella wanted to be wrapped in a shawl, hidden away, asleep, rocked by the march, a baby, a round infant in arms. Magda took Rosa's nipple, and Rosa never stopped walking, a walking cradle. There was not enough milk; sometimes Magda sucked air; then she screamed. Stella was ravenous. Her knees were tumors on sticks, her elbows chicken bones.

Cynthia Ozick was born in New York in 1928 and is an acclaimed novelist, short story writer, essayist, and critic. She comments, "Inventing a secret, then revealing it in the drama of entanglement — this is what ignites the will to write stories...

The secrets that engage me — that sweep me away — are generally secrets of inheritance: how the pear seed becomes a pear tree, for instance, rather than a polar bear."

Rosa did not feel hunger; she felt light, not like someone walking but like someone in a faint, in trance, arrested in a fit, someone who is already a floating angel, alert and seeing everything, but in the air, not there, not touching the road. As if teetering on the tips of her fingernails. She looked into Magda's face through a gap in the shawl: a squirrel in a nest, safe, no one could reach her inside the little house of the shawl's windpings. The face, very round, a pocket mirror of a face: but it was not Rosa's bleak complexion, dark like cholera, it was another kind of face altogether, eyes blue as air, smooth feathers of hair nearly as yellow as the Star sewn into Rosa's coat. You could think she was one of *their* babies.

Rosa, floating, dreamed of giving Magda away in one of the villages. She could leave the line for a minute and push Magda into the hands of any woman on the side of the road. But if she moved out of line they might shoot. And even if she fled the line for half a second and pushed the shawl-bundle at a stranger, would the woman take it? She might be surprised, or afraid; she might drop the shawl, and Magda would fall out and strike her head and die. The little round head. Such a good child, she gave up screaming, and sucked now only for the taste of the drying nipple itself. The neat grip of the tiny gums. One mite of a tooth tip sticking up in the bottom gum, how shiny, an elfin tombstone of white marble gleaming there. Without complaining, Magda relinquished Rosa's teats, first the left, then the right; both were cracked, not a sniff of milk. The duct-crevice extinct, a dead volcano, blind eye, chill hole, so Magda took the corner of the shawl and milked it instead. She sucked and sucked, flooding the threads with wetness. The shawl's good flavour, milk of linen.

It was a magic shawl, it could nourish an infant for three days and three nights. Magda did not die, she stayed alive, although very quiet. A peculiar smell, of cinnamon and almonds, lifted out of her mouth. She held her eyes open every moment, forgetting how to blink or nap, and Rosa and sometimes Stella studied their blueness. On the road they raised one burden of a leg after another and studied Magda's face. "Aryan," Stella said, in a voice grown as thin as a string; and Rose thought how Stella gazed at Magda like a young cannibal. And the time that Stella said "Aryan," it sounded to Rosa as if Stella had really said "Let us devour her."

But Magda lived to walk. She lived that long, but she did not walk very well, partly because she was only fifteen months old, and partly because the spindles of her legs could not hold up her fat belly. It was fat with air, full and round. Rosa gave almost all her food to Magda, Stella gave nothing; Stella was ravenous, a growing child herself, but not growing much. Stella did not menstruate. Rosa did not menstruate. Rosa was ravenous, but also not; she learned from Magda how to drink the taste of a finger in one's mouth. They were in a place without pity, all pity was annihilated in Rosa, she looked at Stella's bones without pity. She was sure that Stella was waiting for Magda to die so she could put her teeth into the little thighs.

Rosa knew Magda was going to die very soon; she should have been dead already, but she had been buried away deep inside the magic shawl, mistaken for the shivering mound of Rosa's breasts; Rosa clung to the shawl as if it covered only herself. No one took it

away from her. Magda was mute. She never cried. Rosa hid her in
the barracks, under the shawl, but she knew that one day someone
would inform; or one day someone, not even Stella, would steal
Magda to eat her. When Magda began to walk, Rosa knew that
Magda was going to die very soon, something would happen. She
was afraid to fall asleep; she slept with the weight of her thigh on
Magda's body; she was afraid she would smother Magda under her
thigh. The weight of Rosa was becoming less and less; Rosa and
Stella were slowly turning into air.

Magda was quiet, but her eyes were horribly alive, like blue
tigers. She watched. Sometimes she laughed — it seemed a laugh,
but how could it be? Magda had never seen anyone laugh. Still,
Magda laughed at her shawl when the wind blew its corners, the bad
wind with pieces of black in it, that made Stella's and Rosa's eyes tear.
Magda's eyes were always clear and tearless. She watched like a tiger.
She guarded her shawl. No one could touch it; only Rosa could touch
it. Stella was not allowed. The shawl was Magda's own baby, her pet,
her little sister. She tangled herself up in it and sucked on one of the
corners when she wanted to be very still.

Then Stella took the shawl away and made Magda die.
Afterward Stella said: "I was cold."

And afterward she was always cold, always. The cold went
into her heart: Rosa saw that Stella's heart was cold. Magda flopped
onward with her little pencil legs scribbling this way and that, in
search of the shawl; the pencils faltered at the barracks opening,
where the light began. Rosa saw and pursued. But already Magda
was in the square outside the barracks, in the jolly light. It was the
roll-call arena. Every morning Rosa had to conceal Magda under the
shawl against a wall of the barracks and go out and stand in the arena
with Stella and hundreds of others, sometimes for hours, and Magda,
deserted, was quiet under the shawl, sucking on her corner. Every
day Magda was silent, and so she did not die. Rosa saw that today
Magda was going to die, and at the same time a fearful joy ran in
Rosa's two palms, her fingers were on fire, she was astonished, febrile:
Magda, in the sunlight, swaying on her pencil legs, was howling.
Ever since the drying up of Rosa's nipples, ever since Magda's last
scream on the road, Magda had been devoid of any syllable; Magda
was a mute. Rosa believed that something had gone wrong with her

vocal cords, with her windpipe, with the cave of her larynx; Magda was defective, without a voice; perhaps she was deaf; there might be something amiss with her intelligence; Magda was dumb. Even the laugh that came when the ash-stippled wind made a clown out of Magda's shawl was only the air-blown showing of her teeth. Even when the lice, head lice and body lice, crazed her so that she became as wild as one of the big rats that plundered the barracks at daybreak looking for carrion, she rubbed and scratched and kicked and bit and rolled without a whimper. But now Magda's mouth was spilling a long viscous rope of clamour.

"Maaaa——"

It was the first noise Magda had ever sent out from her throat since the drying up of Rosa's nipples.

"Maaa...aaa!"

Again! Magda was wavering in the perilous sunlight of the arena, scribbling on such pitiful little bent shins. Rosa saw. She saw that Magda was grieving for the loss of her shawl, she saw that Magda was going to die. A tide of commands hammered in Rosa's nipples: Fetch, get, bring! But she did not know which to go after first, Magda or the shawl. If she jumped out into the arena to snatch Magda up, the howling would not stop, because Magda would still not have the shawl; but if she ran back into the barracks to find the shawl, and if she found it, and if she came after Magda holding it and shaking it, then she would get Magda back, Magda would put the shawl in her mouth and turn dumb again.

Rosa entered the dorm. It was easy to discover the shawl. Stella was heaped under it, asleep in her thin bones. Rosa tore the shawl free and flew — she could fly, she was only air — into the arena. The sunheat murmured of another life, of butterflies in summer. The light was placid, mellow. On the other side of the steel fence, far away, there were green meadows speckled with dandelions and deep-coloured violets; beyond them, even farther, innocent tiger lilies, tall, lifting their orange bonnets. In the barracks they spoke of "flowers," of "rain": excrement, thick turd-braids, and the slow stinking maroon waterfall that slunk down from the upper bunks, the stink mixed with a bitter fatty floating smoke that greased Rosa's skin. She stood for an instant at the margin of the arena. Sometimes the electricity inside the fence would seem to hum; even Stella said it was only an imagining, but Rosa heard real sounds in the wire: grainy sad voices. The farther she was from the fence, the more clearly the voices crowded at her. The lamenting voices strummed so convincingly, so passionately, it was impossible to suspect them of being phantoms. The voices told her to hold up the shawl, high; the voices told her to shake it, to whip with it, to unfurl it like a flag. Rosa lifted, shook, whipped, unfurled. Far off, very far, Magda leaned across her air-fed belly, reaching out with the rods of her arms. She was high up, elevated, riding someone's shoulder. But the shoulder that carried Magda was not coming toward Rosa and the shawl, it was drifting away, the speck of Magda was moving more and more into the smoky distance. Above the shoulder a helmet glinted. The light tapped the helmet and sparkled it into a goblet. Below the helmet a black body like a domino and a pair of black boots hurled themselves in the direction of the electrified fence. The electric voices began to chatter wildly. "Maamaa, maaamaaa," they all hummed together. How far Magda was from Rosa now, across the whole square, past a dozen barracks, all the way on the other side. She was no bigger than a moth.

All at once Magda was swimming through the air. The whole of Magda travelled through loftiness. She looked like a butterfly touching a silver vine. And the moment Magda's feathered round head and her pencil legs and balloonish belly and zigzag arms splashed against the fence, the steel voices went mad in their growling, urging Rosa to run and run to the spot where Magda had fallen from her flight against the electrified fence; but of course Rosa did not obey them. She only stood, because if she ran they would shoot, and if she tried to pick up the sticks of Magda's body they would shoot, and if she let the wolf's screech ascending now through the ladder of her skeleton break out, they would shoot; so she took Magda's shawl and filled her own mouth with it, stuffed it in and stuffed it in, until she was swallowing up the wolf's screech and tasting the cinnamon and almond depth of Magda's saliva; and Rosa drank Magda's shawl until it dried.

Mother

TONGUE

AMY TAN

Amy Tan was born in 1952 in Oakland, California, far from the childhood worlds of her China-born parents. The novel that changed her life was Louise Erdrich's Love Medicine, from which came her belief in her own writing and in the truth as "a multiple story." Consequently, Tan has been sharply critical of her experiences in schools where students' abilities are measured using standardized tests: "They don't measure creativity or 'good' differences; they see differences (only) as deficits."

I AM NOT A SCHOLAR OF ENGLISH OR LITERATURE. I cannot give you much more than personal opinions on the English language and its variations in this country or others.

I am a writer. And by that definition, I am someone who has always loved language. I am fascinated by language in daily life. I spend a great deal of my time thinking about the power of language — the way it can evoke an emotion, a visual image, a complex idea, or a simple truth. Language is the tool of my trade. And I use them all — all the Englishes I grew up with.

Recently, I was made keenly aware of the different Englishes I do use. I was giving a talk to a large group of people, the same talk I had already given to half a dozen other groups. The nature of the talk was about my writing, my life, and my book, *The Joy Luck Club.* The talk was going along well enough, until I remembered one major difference that made the whole talk sound wrong. My mother was in the room, And it was perhaps the first time she had heard me give a lengthy speech, using the kind of English I have never used with her. I was saying things like, "The intersection of my memory upon imagination" and "There is an aspect of my fiction that relates to thus-and-thus" — a speech filled with carefully wrought grammatical phrases, burdened, it suddenly seemed to me, with nominalized forms, past perfect tenses, conditional phrases, all the forms of standard English that I had learned in school and through books, the forms of English I did not use at home with my mother.

Just last week, I was walking down the street with my mother and I again found myself conscious of the English I was using, the English I do use with her. We were talking about the price of new and used furniture and I heard myself saying this: "Not waste money that way." My husband was with us as well, and he didn't notice any switch in my English. And then I realized why. It's because over the twenty years we've been together I've often used that same kind of English with him, and sometimes he even uses it with me. It has become our language of intimacy, a different sort of English that relates to family talk, the language I grew up with.

So you'll have some idea of what this family talk I heard sounds like, I'll quote what my mother said during a recent conversation which I videotaped and then transcribed. During this conversation, my mother was talking about a political gangster in Shanghai who had the same last name as her family's, Du, and how the gangster in his early years wanted to be adopted by her family, which was rich by comparison. Later, the gangster became more powerful, far richer than my mother's family, and one day showed up at my mother's wedding to pay his respects. Here's what she said in part:

"Du Yusong having business like fruit stand. Like off the street kind. He is Du like Du Zong — but not Tsung-ming Island people. The local people call putong, the river east side, he belong to that side local people. That man want to ask Du Zong father take him in like become own family. Du Zong father wasn't look down on him, but didn't take seriously, until that man big like become a mafia. Now important person, very hard to inviting him. Chinese way, came only to show respect, don't stay for dinner. Respect for making big celebration, he shows up. Mean gives lots of respect. Chinese custom. Chinese social life that way. If too important won't have to stay too long. He come to my wedding. I didn't see, I heard it. I gone to boy's side, they have YMCA dinner. Chinese age I was nineteen."

You should know that my mother's expressive command of English belies how much she actually understands. She reads the Forbes report, listens to *Wall Street Week,* converses daily with her stockbroker, reads all of Shirley MacLaine's books with ease — all kinds of things I can't begin to understand. Yet some of my friends tell me they understand 50 percent of what my mother says. Some say they understand 80 to 90 percent. Some say they understand none of it, as if she were speaking pure Chinese. But to me, my mother's English is perfectly clear, perfectly natural. It's my mother tongue. Her language, as I hear it, is vivid, direct, full of observation and imagery. That was the language that helped shape the way I saw things, expressed things, made sense of the world.

Lately, I've been giving more thought to the kind of English my mother speaks. Like others, I have described it to people as "broken" or "fractured" English. But I wince when I say that. It has always bothered me that I can think of no way to describe it other than "broken," as if it were damaged and needed to be fixed, as if it lacked a certain wholeness and soundness. I've heard other terms used, "limited English," for example. But they seem just as bad, as if everything is limited, including people's perceptions of the limited English speaker.

I know this for a fact, because when I was growing up, my mother's "limited" English limited my perception of her. I was ashamed of her English. I believed that her English reflected the quality of what she had to say. That is, because she expressed them imperfectly, her thoughts wree imperfect. And I had plenty of empirical evidence to support me: the fact that people in department stores, at banks, and at restaurants did not take her seriously, did not give her good service, pretended not to understand her, or even acted as if they did not hear her.

My mother has long realized the limitations of her English as well. When I was fifteen, she used to have me call people on the phone to pretend I was she. In this guise, I was forced to ask for information or even to complain and yell at people who had been rude to her. One time it was a call to her stockbroker in New York. She had cashed out her small portfolio and it just so happened we were going to New York the next week, our very first trip outside California. I had to get on the phone and say in an adolescent voice that was not very convincing, "This is Mrs. Tan."

And my mother was standing in the back whispering loudly, "Why he don't send me check, already two weeks late. So mad he lie to me, losing me money."

And then I said in perfect English, "Yes, I'm getting rather concerned. You had agreed to send the check two weeks ago, but it hasn't arrived."

Then she began to talk more loudly. "What he want, I come to New York tell him in front of his boss, you cheating me?" And I was trying to calm her down, make her be quiet, while telling the stockbroker, "I can't tolerate any more excuses. If I don't receive the check immediately, I am going to have to speak to your manager when I'm in New York next week." And sure enough, the following week there we were in front of this astonished stockbroker, and I was sitting there red-faced and quiet, and my mother, the real Mrs. Tan, was shouting at his boss in her impeccable broken English.

We used a similar routine just five days ago, for a situation that was far less humorous. My mother had gone to the hospital for an appointment, to find out about a benign brain tumor a CAT scan had revealed a month ago. she said she had spoken very good English, her best English, no mistakes. Still, she said, the hospital did not apologize when they said they had lost the CAT scan and she had come for nothing. She said they did not seem to have any sympathy when she told them she was anxious to know the exact diagnosis, since her husband and son had both died of brain tumors. She said they would not give her any more information until the next time and she would have to make another appointment for that. So she said she would not leave until the doctor called her daughter. She wouldn't budge. And when the doctor finally called her daughter, me, who spoke in perfect English — lo and behold — we had assurances the CAT scan would be found, promises that a conference call on Monday would be held, and apologies for any suffering my mother had gone through for a most regrettable mistake.

I think my mother's English almost had an effect on limiting my possibilities in life as well. Sociologists and linguists probably will tell you that a person's developing language skills are more influenced by peers. But I do think that the language spoken in the family, especially in immigrant families which are more insular, plays a large role in shaping the language of the child. And I believe that it affected my results on achievement tests, IQ tests, and the SAT. While my English skills were never judged as poor, compared to math, English could not be considered my strong suit. In grade school I did moderately well, getting perhaps B's, sometimes B-pluses, in English and scoring perhaps in the sixtieth or seventieth percentile on achievement tests. But those scores were not good enough to override the opinion that my true abilities lay in math and science, because in those areas I achieved A's and scored in the ninetieth percentile or higher.

This was understandable. Math is precise; there is only one correct answer. Whereas, for me at least, the answers on English tests were always a judgement call, a matter of opinion and personal experience. Those tests were constructed around items like fill-in-the-blank sentence completion, such as, "Even though Tom was _____, Mary thought he was _____." And the correct answer always seemed to be the most bland combinations of thoughts, for example, "Even though Tom was shy, Mary thought he was charming," with the grammatical structure "even though" limiting the correct answer to some sort of semantic opposites, so you wouldn't get answers like, "Even though Tom was foolish, Mary thought he was ridiculous." Well, according to my mother, there were very few limitations as to what Tom could have been and what Mary might have thought of him. So I never did well on tests like that.

The same was true with word analogies, pairs of words in which you were supposed to find some sort of logical, semantic relationship — for example, "Sunset is to nightfall as _____ is to _____." And here you would be presented with a list of four possible pairs, one of which showed the same kind of relationship: red is to stoplight, bus is to arrival, chills is to fever, yawn is to boring. Well, I could never think that way. I knew what the tests were asking, but I could not block out of my mind the images already created by the first pair, "sunset is to nightfall" — and I would see a burst of colours against a darkening sky, the moon, rising, the lowering of a curtain of stars. And all the other pairs of words — red, bus, stoplight, boring— just threw up a mass of confusing images, making it impossible for me to sort out something as logical as saying "A sunset precedes nightfall" is the same as "A chill precedes a fever." The only way I would have gotten that answer right would have been to imagine an associative situation, for example, my being disobedient and staying out past sunset, catching a chill at night, which turns into feverish pneumonia as punishment, which indeed did happen to me.

I have been thinking about all this lately, about my mother's English, about achievement tests. Because lately I've been asked, as a writer, why there are not more Asian Americans represented in American literature. Why are there few Asian Americans enrolled in creative writing programs? Why do so many Chinese students go into engineering? Well, these are broad sociological questions I can't begin to answer. But I have noticed in surveys — in fact, just last week — that Asian students, as a whole, always do significantly better on math achievement tests than in English, And this makes me think that there are other Asian-American students whose English spoken in the home might also be described as "broken" or "limited," And perhaps they also have teachers who are steering them away from writing and into math and science, which is what happened to me,

Fortunately, I happen to be rebellious in nature and enjoy the challenge of disproving assumptions made about me. I became an English major my first year in college, after being enrolled as pre-med. I started writing nonfiction as a freelancer the week after I was told by my former boss that writing was my worst skill and I should hone my talents toward account management.

But it wasn't until 1985 that I finally began to write fiction. And at first I wrote using what I thought to be wittily crafted sentences, sentences that would finally prove I had mastery over the English language. Here's an example from the first draft of a story that later made its way into *The Joy Luck Club*, but without this line: "That was my mental quandary in its nascent state." A terrible line, which I can barely pronounce.

Fortunately, for reasons I won't get into today, I later decided I should envision a reader for the stories I would write. And the reader I decided upon was my mother, because these were stories about mothers. So with this reader in mind — and in fact she did read my early drafts — I began to write stories using all the Englishes I grew up with: the English I spoke to my mother, which for lack of a better term might be described as "simple;" the English she used with me, which for lack of a better term might be described as "broken," my translation of her Chinese, which could certainly be described as "watered down;" and what I imagined to be her translation of her Chinese if she could speak in perfect English, her internal language, and for that I sought to preserve the essence, but neither an English nor a Chinese structure. I wanted to capture what language ability tests can never reveal: her intent, her passion, her imagery, the rhythms of her speech and the nature of her thoughts.

Apart from what any critic had to say about my writing, I knew I had succeeded where it counted when my mother finished reading my book and gave me her verdict: "So easy to read."

NIGHT

TATYANA TOLSTAYA

IN THE MORNING ALEXEI PETROVICH'S Mommy yawns very very loudly; hurrah, up and at 'em! The new morning sprays into the window; the cactuses shine, the curtain trembles; the gates of the nighttime kingdom have slammed shut; the dragons, toadstools, and terrifying dwarfs have vanished underground again, life triumphs and heralds trumpet: A new day! A new day! Too-roo-roo-roo-oo-oo-oo!

Mommy scratches her balding head very very fast and tosses her bluish feet off the high bedstead: let them dangle and think a while about having to drag around all day the two hundred and ninety pounds Mommy has amassed in her eighty years

Tatyana Tolstaya was born in Leningrad in 1951 and is the great grandniece of the classic Russian writer, Leo Tolstoy. She writes, "I am inspired by Russian and European folklore, the Old Testament, and crazy contemporary life." Her Soviet publishers excluded "Night" from a published collection of her stories, perhaps because they were uncomfortable with the protagonist who is a mentally handicapped adult, but the story is one of her favourites.

Alexei Petrovich's eyes are wide open by now. Sleep streams gently off his body; the last raven flies off dreamily into the gloom; the nighttime guests have finished gathering up their spectral, ambiguous props and interrupted their performance until the next time. A light draft sweetly fans Alexei Petrovich's bald spot, now and then unshaven stubble prickles his palms. Is it time to get up? Mommy will see to it. Mommy is such a big, booming, capacious woman, and Alexei Petrovich is little. Mommy knows the ropes and gets wherever she wants to. Mommy is all-powerful. What she says, goes. And he is a late child, a little lump, nature's blunder, a fallow patch, a soap sliver, a cockle, an empty husk destined for burning that by chance ended up among its sound brethren when the Sower lavishly broadcast the full-blooded seeds of life over the earth.

Can he get up yet or is it early? Don't give a peep. Mommy

is completing her morning ritual: she trumpets into a handkerchief, pulls clinging stockings up the columns of her legs, and fastens them under her swollen knees with little ringlets of white rubberbands. She raises a linen frame with fifteen little buttons onto her enormous bosom; it's probably hard to fasten them from behind. A gray chignon is fixed to Mommy's zenith; refreshed teeth flutter out of the clean night glass and shake themselves off. Mommy's facade is covered by a white fluted dickey, and a coarse navy casing goes over the entire majestic edifice, concealing laces down the back, wrong sides, rear formations, service stairs and emergency exits. The palace is erected.

Everything you do is fine, Mommy. It's all right.

All the Men and Women in the communal apartment are already awake, bustling about and talking. Doors bang, water gurgles, tinkling noises come through the wall. The ship of morning has left the slip, it slices through the blue water, wind fills its sails, and well-dressed travellers laugh and exchange remarks on the deck. What lands lie ahead? Mommy is at the helm, Mommy is on the bridge; from the tip of the mast Mommy peers into the shining ripples.

"Alexei, get up! Shave, brush your teeth, and wash your ears! Take a clean towel. Screw the cap on the toothpaste! Don't forget to flush. And don't touch anything in there, do you hear me?"

Fine, fine, Mommy. That's how you always say the right thing. That's the way it all suddenly makes sense, the way the horizons are flung open and it's safe to sail with an experienced pilot! The ancient coloured maps are unrolled, the route is drawn in a red dotted line and bright, easy-to-understand pictures mark all the dangers: here's a menacing lion, and on that shore there's a rhinoceros; here a whale spouts like a toylike fountain, and over there is the most dangerous of all, the big-eyed, long-tailed Sea Maiden, slippery, deadly, and enticing.

Now Alexei Petrovich will wash and tidy himself; Mommy will come to check that he hasn't made a mess in there, or else the neighbours will make trouble again; and then something yummy to eat! What has Mommy made there today? You have to fight your way through the kitchen to get to the bathroom. Old women grumble over hot stoves, brewing poison in small buckets, adding the roots of frightful herbs and following Alexei Petrovich with their bad glances. Mommy! Don't let them hurt me!

He splashed a bit on the floor. Oy.

There's a crowd in the hall already; the Men and Women are leaving, they make noise, check their keys and purses

The corner door with the frosted panes is wide open; the impudent Sea Maiden stands on the threshold, smirking and winking at Alexei Petrovich; she is all tilted; she blazes with Tobacco and has her leg thrust out; she has spread her net — wouldn't you like to get caught, eh? But Mommy will save him, she's already rushing up like a steam engine, red wheels clacking, hooting: Out of the way!

"Brazen hussy! Go away, I tell you! It's not enough that you ... and to a sick man yet!"

"Ha-ha-ha!" The Sea Maiden isn't scared.

Quick — into the room. He's safe. Phoo-oo-oohh... Women — it's very frightening. It's not clear why, but it's very upsetting. They walk by — they smell so nice ... and they have — Legs. There are lots of them on the street, and in every house, that one, and that one, and this one, behind every door; they're hiding, doing something, bending down, rummaging around, tittering behind their hands; they know, but they won't tell Alexei Petrovich. So he'll sit down at the table and think about Women. One day Mommy took him to a beach in the country where there were a lot of them. There was one there, kind of ... a wavy kind of fairy ... like a little dog... Alexei Petrovich liked her. He went up close and started looking.

"Well, anything you haven't seen?" cried the fairy. "Blow, get out of here, you retard!"

Mommy came in carrying a sizzling saucepan. He peeked. In it were rosy little nozzles of sausage. He was glad. Mommy loads up his plate, moves things around, wipes up. The knife breaks loose from his fingers and strikes sharply somewhere to one side, cutting the oilcloth.

"Your hand, pick up the sausage in your hand!"

Oh, Mommy, guiding star! You're pure gold! You'll take care of everything, wise woman, you'll untangle all the knots! You knock down all the dark corners, all the labyrinths of the incomprehensible, impassable world with your mighty arm; you sweep away all the barriers — now the ground is flat and level. Be bold, take another step! But farther on — there are more wind-fallen trees.

Alexei Petrovich has his world, the real one, in his head. There everything is possible, And this one, the outer one, is wicked and wrong. And it's very hard to keep in mind what's good and what's bad. Here they've made arrangements and come to agreements, they've written Rules, horribly complicated ones. They've learned them, they have good memories. But it's hard for him to live by other people's Rules.

Mommy poured coffee. Coffee has a Smell. Drink it, and it shifts to you. Why aren't you permitted to stick out your lips, squint your eyes to your mouth, and sniff yourself? Wait until Mommy turns her back.

"Alexei, behave yourself!"

After breakfast they cleared the table, set out glue and cardboard, laid out scissors, and tied a napkin around Alexei Petrovich: he was going to glue little boxes. When he finished a hundred, they took them to the pharmacy. That brought in a little money. Alexei Petrovich really loves the boxes, he's sorry to part with them. He wants to hide some unnoticed, to keep at least a few for himself, but Mommy keeps a sharp watch and takes them away.

And afterwards strangers carry them out of the pharmacy, eat little white globes from them, and tear up the boxes and throw them away! They throw them right in the wastebin, and not only there — in their own apartment, in the kitchen, in the garbage pail he sees a ripped and defiled little box with a cigarette butt inside! Then a terrible, black rage overwhelms Alexei Petrovich, his eyes glitter, he gushes saliva and forgets words, fiery spots jump before his eyes, and he's ready to strangle and tear someone to pieces. Who did it? Who dared to do it? Just get out! He rolls up his sleeves: where is he? Mommy comes running, placates and leads off the infuriated Alexei Petrovich, takes away the knife, and tears the hammer from his convulsively clutching fingers. The Men and Women are afraid then and sit quietly, taking refuge in their rooms.

The sun moved across to the other window. Alexei Petrovich's work is done. Mommy has fallen asleep in the armchair, she snores, her cheeks gurgling, and hisses: Psht-sht-sht... Alexei Petrovich very very quietly takes two boxes and cau-autiously, on tiptoe, tup tup tuppity, goes over to the bed and car-r-refully puts them under the pillow. At night he will get them out and sniff. How nice the glue smells! Soft, sour, muffled, like the letter "f."

Mommy woke up, it was time for a walk. Down the stairs, never in the elevator — you couldn't shut Alexei Petrovich in the elevator, he would struggle and squeal like a rabbit. Why can't you understand — they pull, pull on your legs and drag you down!

Mommy sails out ahead, exchanging nods with acquaintances. Today we are delivering the boxes: it's unpleasant. Alexei Petrovich deliberately trips over his own feet: he doesn't want to go to the pharmacy.

"Alexei, put your tongue in!"

Sunset fell behind the tall buildings. Golden panes burned just under the roof. Special people live there, not like us: they fly like white doves, fluttering from balcony to balcony. A smooth feathery little breast, a human face — if a bird like that should land on your railing and bend its head and begin to coo, you would stare into its eyes, forget the human tongue, and begin to trill like a bird and jump with shaggy little legs along your iron perch.

Below the horizon, below the plate of the earth, gigantic wheels had begun to turn, monstrous belt drives revolved, and cogwheels drew the sun down and the moon up. The day was tired, he had folded his white wings and flown off westward; large in his roomy garments, he waved his sleeve and released the stars, he blessed those walking on the cooling earth: Until we meet again, until we meet again, I'll come back tomorrow.

On the corner they're selling ice cream. He really wants some! The Men and Women — but especially the Women — thrust bits of money into a square little window and receive a frozen crunchy goblet. They laugh, they throw the round sticky papers on the ground or stick them to the wall, they open their mouths wide and lick the sweet spiky cold with their red tongues.

"Mommy, ice cream!"

"You can't have any. You have a sore throat."

If he couldn't, then he couldn't. But he really wanted some! It was awful how much he wanted some! If only he had bits of money like the other Men and Women, silver and shiny; or a yellow paper that smelled of bread — they took those in the square window, too.

Oy, oy, oy, how he wanted some, they all could have it, they all got it

"Alexei, stop turning your head!"

Mommy knows best. I'll listen to Mommy. Only she knows the true path through the wilds of the world. But if Mommy were to turn her back ... Pushkin Square.

"Mommy, is Pushkin a writer?"

"Yes."

"I'm going to be a writer, too."

Of course you are. If you want to — you will be."

And why not? If he wants to — he will. He'll take a piece of paper and a pencil, and he'll be a writer. That's all, it's decided! He'll be a writer. That's fine.

In the evenings Mommy sits in her roomy armchair, lowers glasses onto her nose, and reads in a deep voice:

> Storm the sky with darkness cowls,
> Snow in eddies whirls and sweeps;
> Now like a feral beast it howls,
> Now like a little child it weeps...

Alexei Petrovich really likes that a lot! He laughs broadly, baring yellow teeth, he rejoices and stamps his feet.

> Now like a feral beast it howls,
> Now like a little child it weeps...

Just like that the words reach the end — and then turn back, and go on again — and then turn back again.

> Stor mthes kywi thdar knessc owls
> Sno win ed dies whir Isan dswe eps
> Nowl ikeaf er albea sti thowls
> Nowl ikeal itt lech il ditwe eps

That's great! Here's how it howls: oo-oo-oo-oo-oo!

"Hush, hush, Alexei, calm down!"

The sky is all covered with stars. Alexei Petrovich knows

them: little shiny beads, suspended on their own in the black void. As Alexei Petrovich lies in bed trying to get to sleep, his legs start to grow down, down, all by themselves, and his head up, up to the black dome, always up, and it rocks back and forth like a treetop in a storm, and the stars scratch his skull. But the second Alexei Petrovich, the inner one, keeps shrinking and shrinking; he contracts, disappearing into a poppyseed, into the sharp end of a needle, into a little microbe, into nothing and, if he is not stopped, he will vanish into it altogether. But the outer, gigantic Alexei Petrovich sways like a lodgepole pine, grows, and strikes his bald spot sharply against the night dome — he won't let the little one disappear into a point. And these two Alexei Petroviches are one and the same. And that's sensible, that's right.

At home Mommy undresses, demolishes her daytime body, puts on a red robe, and becomes simpler, warmer, and easier to understand. Alexei Petrovich wants to jump into Mommy's arms! What nonsense! Mommy goes off to the kitchen. For some reason she's away for a long time. Alexei Petrovich checked that the little boxes were still in place, sniffed the glue, and took a risk — he went out into the hall. The corner door, where at night the Sea Maiden's guests titter, is ajar. He can see a white bed. But where's Mommy? Maybe she's there? Alexei Petrovich peeks cautiously through the crack. Nobody there. Maybe Mommy is hidden behind the wardrobe? Should he go in? The room is empty. On the Sea Maiden's table are open cans, bread, and a pickle with a bite out of it. And something else — a yellow paper and little round silver bits. Money! Take the money, rush down the dark stairs, into the labyrinths of the streets, search out the square window, there they'll give you a sweet cold little glass!

Alexei Petrovich snatches, jingles, overturns, runs, bangs the door, breathes noisily and hastily, stumbles. The street. Darkness. Which way? That way? Or this? What's that in his fist? Put your hand in your pocket. No, it glows through anyway. Somebody else's money! He took somebody else's money! Passersby turn around and whisper to one another: "He took somebody else's money!" People pressed to the windows and nudged each other: let me see! Where is he? Over there! He's got money! Ah-ha, you took it? Alexei Petrovich runs into the darkness. Clink, clink, clink, clink — it's the money in his pocket. The entire city pours out into the street. Shutters are flung open. Arms jab from every window, eyes glitter, long red tongues are thrust out: "He took money! Loose the dogs!" Fire engines howl, hoses uncoil: where is he? Over there! Go get him! Alexei Petrovich thrashes about in panic! Throw it away, rip it from your hands, get rid of it, get rid of it, that's it, there! Use your foot, your foot! Tr-r-ram-m-ple the bits! That's the way.... All of them .... They aren't breathing. They're silent now. The glow's gone. He wipes his face. That's it. Which way now? Night. It smells. Where's Mommy? Night. In gateways wolves stand in black ranks, waiting. I'll walk backwards, that'll fool them. That's good. It's stifling. I'll undo my buttons. I'll unbutton everything. Good. Now? Women with Legs went past. They turned around. They chortled. Ah, so, so-o-o-! Wh-a-at? Me? I'm a wolf! I'm walking backwards! Aha, that's scared you, has it? Now I'll catch up and pounce on you — we'll see what sort of Legs you have! He rushed at them. A scream. A-a-a-ah! A blow. Don't hit me! A blow. Men smelling of Tobacco hit him in the stomach, in the teeth! Don't! ... Oh, leave him alone — can't you see .... They walked away.

Alexei Petrovich propped himself against a downspout, spat black stuff, and whimpered. Poor little fellow, all alone, lost on the street, you came into this world by mistake! Get out of here, it's not for you! Alexei Petrovich cries with a loud barking noise, raising his disfigured face to the stars.

Mommy, Mommy, where are you? Mommy, the way is dark, the voices are silent, and the paths lead into trackless swamps! Mommy, your only, your best beloved, long awaited child, borne in suffering, is weeping, he's dying!...

Mommy comes running, Mommy gasps for breath, stretches out her arms, shouts, seizes him, clasps him to her breasts, runs her hands over him, kisses him. Mommy sobs — I found him, I found him!

Mommy leads Alexei Petrovich by the bridle into the warm burrow, the soft nest, under her white wing.

His swollen face is washed! Alexei Petrovich sits snivelling at the table with a napkin tied around him.

"Do you want a nice soft-boiled egg? Soft, all runny?"

Alexei Petrovich nods: Yes, I do. The wall clock ticks.

Peace. Delicious hot milk, mellow, like the letter "n." Something becomes clear in his head. Yes! He wanted...

"Mommy, give me paper and a pencil. Quick! I'm going to be a writer!"

"Oh Lord! Misery mine! What on earth do you .... Well, don't cry, calm down, I'll get them; hold on, you need to blow your nose.

White paper and a sharp pencil. Quick, quick, before he forgets! He knows it all, he's made sense of the world, made sense of the Rules, he's grasped the secret connection among events, grasped the laws that link millions of fragments of disparate things! Lightning flashes in Alexei Petrovich's brain! He frets, grumbles, snatches a sheet of paper, shoves glasses aside with his elbow, and, amazed himself at his joyful regeneration, hastily, in large letters, records his newly found truth: "Night. Night. Night. Night. Night. Night. Night. Night. Night. Night."

ATTRACTA

WILLIAM TREVOR

William Trevor was born in Mitchelstown, County Cork, Ireland in 1928. Of the characters who people his invented landscapes, he writes: "There are no ordinary people and there are no villains or heroes. When I was a sculptor I was very aware of the ordinariness of wood and clay and what can come of them. That is the same with my writing."

ATTRACTA READ ABOUT PENELOPE VADE IN A newspaper, an item that upset her. It caused her to wonder if all her life as a teacher she'd been saying the wrong things to the children in her care. It saddened her when she thought about the faces that had passed through her schoolroom, ever since 1937. She began to feel she should have told them about herself.

In the schoolroom Attracta taught the sixteen Protestant children of the town. The numbers had been sometimes greater in the past, and often fewer; sixteen was an average, a number she found easy to manage when divided into the four classes that the different ages demanded. The room was large, the desks arranged in groups, discipline had never been a problem. The country children brought sandwiches for lunch, the children of the town went home at midday. Attracta went home herself, to the house in North Street which she'd inherited from her Aunt Emmeline and where now she lived alone. She possessed an old blue Morris Minor but she did not often drive it to and from her schoolroom, preferring to make the journey on foot in order to get fresh air and exercise. She was a familiar figure, the Protestant teacher with her basket of groceries or exercise-books. She had never married, though twice she'd been proposed to: by an exchange clerk in the Provincial Bank and by an English visitor who'd once spent the summer in the area with his parents. All that was a long time ago now, for Attracta was sixty-one. Her predecessor in the schoolroom, Mr. Ayrie, hadn't retired until he was over seventy. She had always assumed she'd emulate him in that.

Looking back on it, Attracta didn't regret that she had not

married. She hadn't much cared for either of the men who'd proposed to her and she didn't mind being alone at sixty-one in her house in North Street. She regularly went to church, she had friends among the people who had been her pupils in the past. Now and again in the holidays she drove her Morris Minor to Cork for a day's shopping and possibly a visit to the Savoy or the Pavilion, although the films they offered were not as good as they'd been in the past. Being on her own was something she'd always known, having been both an only child and an orphan. There'd been tragedy in her life but she considered that she had not suffered. People had been good to her.

English Girl's Suicide in Belfast the headline about Penelope Vade said, and below it there was a photograph, a girl with a slightly crooked smile and freckled cheeks. There was a photograph of her husband in army uniform, taken a few weeks before his death, and of the house in Belfast in which she had later rented a flat. *From the marks of blood on carpets and rugs, the item said, it is deduced that Mrs. Vade dragged herself across the floors of two rooms. She appears repeatedly to have fainted before she reached a bottle of aspirins in a kitchen cupboard.* She had been twenty-three at the time of her death.

It was Penelope Vade's desire to make some kind of gesture, a gesture of courage and perhaps anger, that had caused her to leave her parents' home in Haslemere and to go to Belfast. Her husband, an army officer, had been murdered in Belfast; he'd been decapitated as well. His head, wrapped in cotton-wool to absorb the ooze of blood, secured within a plastic bag and packed in a biscuit-tin, had been opened by her in Haslemere. She hadn't known that he was dead before his dead eyes stared into hers.

Her gesture was her mourning of him. She went to Belfast to join the Women's Peace Movement, to make the point that somehow neither he nor she had been defeated. But her gesture, publicly reported, had incensed the men who'd gone to the trouble of killing him. One after another, seven of them had committed acts of rape on her. It was after that that she had killed herself.

A fortnight after Attracta had first read the newspaper item it still upset her. It haunted her, and she knew why it did, though only imprecisely. Alone at night, almost catching her unawares, scenes from the tragedy established themselves in her mind: the opening of the biscuit-box, the smell of death, the eyes, blood turning brown.

As if at a macabre slide-show, the scene would change: before people had wondered about her whereabouts Penelope Vade had been dead for four days; mice had left droppings on her body.

One afternoon, in order to think the matter over in peace and quiet, Attracta drove her Morris Minor to the sea at Cedarstrand, eight miles from the town. She clambered from the strand up to the headland and paused there, gazing down into the bay, at the solitary island it held. No one had ever lived on the island because its smallness would have made a self-supporting existence impossible. When she'd been growing up she'd often wondered what it would be like to live alone on the rocky fastness, in a wooden hut or a cottage built of stones. Not very agreeable, she'd thought, for she'd always been sociable. She thought it again as she turned abruptly from the sea and followed a path inland through wiry purple heather.

Two fishermen, approaching her on the path, recognized her as the Protestant teacher from the town eight miles away and stood aside for her to pass. She was thinking that nothing she might ever have said in her schoolroom could possibly have prevented the death of a girl in a city two hundred miles away. Yet in a way it seemed ridiculous that for so long she had been relating the details of Cromwell's desecration and the laws of Pythagoras, when she should have been talking about Mr. Devereux and Geraldine Carey. And it was Mr. Purce she should have recalled instead of the Battle of the Boyne.

The fishermen spoke to her as she passed them by but she didn't reply. It surprised them that she didn't, for they hadn't heard that the Protestant teacher had recently become deaf or odd. Just old, they supposed, as they watched her progressing slowly: an upright figure, spare and seeming fragile, a certain stiffness in her movement.

What made Attracta feel close to the girl in the newspaper item was the tragedy in her own life: the death of her mother and father when she was three. Her parents had gone away, she had been told, and at first she had wept miserably and would not be comforted. But as days passed into weeks, and weeks into months, this unhappiness gradually left her. She ceased to ask about her parents and became used to living in her Aunt Emmeline's house in North Street. In time she no longer remembered the morning she'd woken up in this house in a bed that was strange to her; nor could she recollect her parents' faces. She grew up assuming they were no longer alive and when once she voiced this assumption her aunt did not contradict it. It wasn't until later in her childhood, when she was eleven, that she learnt the details of the tragedy from Mr. Purce, a small man in a hard black hat, who was often to be seen on the streets of the town. He was one of the people she noticed in her childhood, like the elderly beggar-woman called Limerick Nancy and the wild-looking builder's labourer who could walk a hundred miles without stopping, who never wore a jersey or a coat over his open shirt even on the coldest winter days. There were other people too: priests going for a walk in pairs, out along the road that led to the golf-course and to Cedarstrand by the longer route. Strolling through the afternoon sunshine there were nuns in pairs also, and there was Redmond the solicitor hurrying about with his business papers, and Father Quinn on his bicycle. At night there were the red-fleshed country bachelors tipsily smiling through cigarette smoke, lips glistening in the street-light outside Keogh's public house. At all times of day, at all the town's corners, the children of the poor waited for nothing in particular.

The town was everything in Attracta's childhood, and only some of it had changed in the fifty years that had passed. Without nostalgia she remembered now the horses and carts with milk churns for the creamery, slowly progressing on narrow streets between colour-washed houses. On fair-days the pavements had been slithery with dung, and on fair-days they still were. Farmers stood by their animals, their shirts clean for the occasion, a stud at their throats, without collar or tie. Dogs slouched in a manner that was characteristic of the dogs of the town; there was a smell of stout and sawdust. In her childhood there had been O'Mara's Picture House, dour grey cement encasing the dreamland of Fred Astaire and Ginger Rogers. Built with pride in 1929, O'Mara's was a ruin now.

Within the world of the town there was for Attracta a smaller, Protestant world. Behind green railings there was Mr. Ayrie's Protestant schoolroom. There was the Church of Ireland church, with its dusty flags of another age, and Archdeacon Flower's prayers for the English royal family. There were the Sunday-school classes of Mr. and

Mrs. Bell, and the patience of her aunt, which seemed like a Protestant thing also — the Protestant duty of a woman who had never expected to find herself looking after a child. There was Mr. Devereux, a Protestant who never went to church.

No one in the town, not even her aunt, was kinder to Attracta than Mr. Devereux. On her birthday he came himself to the house in North Street with a present carefully wrapped, a dolls' house once, so big he'd had to ask the man next door to help him out of the dickey of his motor-car with it. At Christmas he had a Christmas tree in his house, and other children in the town, her friends from school, were invited to a party. Every Saturday she spent the afternoon with him, eating his housekeeper's delicious orange cake for tea and sticking stamps into the album he'd given her, listening to the gramophone in the room he called his office. He loved getting a huge fire going in his office, banking up the coals so that they'd glow and redden her cheeks. In summer he sat in his back garden with her, sometimes reading *Coral Island* aloud. He made her run away to the raspberry canes and come back with a punnet of fruit which they'd have at suppertime. He was different from her aunt and from Mr. Ayrie and Archdeacon Flower. He smelt of the tobacco he smoked in his pipe. He wore tweed suits and a striped shirt with a white celluloid collar, and patterned brown shoes which Attracta greatly admired. His tie matched the tweed of his suit, a gold watch dangled from the lapel of his jacket into his top pocket. He was by trade a grain merchant.

His house was quiet and always a little mysterious. The drawing-room, full of looming furniture, was dark in the daytime. Behind layers of curtains that hung to the ground blue blinds obscured the greater part of the light: Sunshine would damage the furniture, Mr. Devereux's housekeeper used to say. On a summer's afternoon this woman would light a paraffin lamp so that she could polish the mahogany surfaces of the tables and the grand piano. Her name was Geraldine Carey: she added to the house's mystery.

Mr. Devereux's smile was slow. There was a laziness about it, both in its leisurely arrival and the way it lingered. His eyes had a weary look, quite out of keeping with all the efforts he made to promote his friendship with Attracta and her aunt. Yet the efforts seemed natural to Attracta, as were the efforts of Geraldine Carey, who was the quietest person Attracta had ever met. She spoke in a voice that it was often hard to hear. Her hair was as black as coal, drawn back from her face and arranged in a coiled bun at the back of her head. Her eyes were startlingly alive, seeming to be black also, often cast down. She had the kind of beauty that Attracta would like one day to possess herself, but knew she would not. Geraldine Carey was like a nun because of the dark clothes she wore, and she had a nun's piety. In the town it was said she couldn't go to mass often enough. "Why weren't you a nun, Geraldine?" Attracta asked her once, watching her making bread in her big, cool kitchen. The habit would have suited her, she added, already imagining the housekeeper's face framed by the coif, and the black voluminous skirts. But Geraldine Carey replied that she'd never heard God calling her. "Only the good are called," she said.

There'd been a time, faintly remembered by Attracta, when her Aunt Emmeline hadn't been well disposed towards Mr. Devereux and Geraldine Carey. There'd been suspicion of some kind, a frowning over presents he brought, an agitation whenever Attracta was invited to tea. Because of her own excitement over the presents and the invitations Attracta hadn't paid much attention to the nature of her aunt's concern, and looking back on it years later could only speculate. Her Aunt Emmeline was a precise person, a tall woman who had never married, reputed to be delicate. Her house in North Street, very different from Mr. Devereux's, reflected her: it was neat as a new pin, full of light, the windows of its small rooms invariably open at the top to let in fresh air. The fan-light above the hall-door was always gleaming, filling the hall with morning sunlight. Attracta's Aunt Emmeline had a fear of darkness, of damp clothes and wet feet, and rain falling on the head. She worried about lots of things.

Clearly she had worried about Mr. Devereux. There was an occasion when Archdeacon Flower had been specially invited to tea, when Attracta had listened at the sitting-room door because she'd sensed from her flustered manner that something important was to be discussed. "Oh, have no worry in that direction at all," she heard the Archdeacon say. "Gentle as a lamb that man's become." Her aunt asked a question Attracta could not hear because of the sound of a tea-cup being replaced on a saucer. "He's doing the best he can," the Archdeacon continued, "according to his lights." Her aunt mentioned Geraldine Carey, and again the Archdeacon reassured her. "Bygones are bygones," he said. "Isn't it a remarkable thing when a man gets caught in his own snare?" He commented on the quality of her aunt's fruitcake, and then said that everyone should be charitably disposed towards Mr. Devereux and Geraldine Carey. He believed, he said, that that was God's wish.

After that, slowly over the years, Attracta's aunt began to think more highly of Mr. Devereux, until in the end there was no one in the entire town, with the possible exception of Archdeacon Flower, whom she held in greater esteem. Once when Phelan the coal merchant insisted that she hadn't paid for half a ton of coal and she recollected perfectly giving the money to the man who'd delivered it, Mr. Devereux had come to her aid. "A right old devil, Phelan is," Attracta heard him saying in the hall, and that was the end her aunt had ever heard of the matter. On Saturday evenings, having kept Attracta company on her walk home, Mr. Devereux might remain for a little while in the house in North Street. He sometimes brought lettuces or cuttings with him, or tomatoes or strawberries. He would take a glass of sherry in the trim little sitting room with its delicate inlaid chairs that matched the delicacy of Attracta's aunt. Often he'd still be there, taking a second glass, when Attracta came down to say goodnight. Her aunt's cat Diggory liked to climb up on to his knees, and as if in respect of some kind Mr. Devereux never lit his pipe. He and her aunt would converse in low voices and generally they'd cease when Attracta entered the room. She would kiss him good-night after she'd kissed her aunt. She imagined it was what having a father was like.

At the town's approximate centre there stood a grey woman on a pedestal, a statue of the Maid of Erin. It was here, only yards from this monument, that Mr. Purce told Attracta the truth about her parents' death, when she was eleven. She'd always had the feeling that Mr. Purce wanted to speak to her, even that he was waiting until she could understand what it was he had to say. He was a man people didn't much like; he'd settled in the town, having come there from somewhere else. He was a clerk in the court-house.

"There's a place I know where there's greenfinches," he said, as if introducing himself. "Ten nests of them, maybe twelve, maybe more. D'you understand me, Attracta? Would you like me to show you?"

She was on her way home from school. She had to get back to do her homework, she said to Mr.Purce. She didn't want to go looking for greenfinches with him.

"Did Devereux tell you not to talk to Mr. Purce?" he said, and she shook her head. As far as she could remember, Mr. Devereux had never mentioned Mr. Purce. "I see you in church," Mr. Purce said.

She had seen him too, sitting in the front, over on the left-hand side. Her aunt had often remarked that the day Mr. Purce didn't go to church it would be a miracle. It was like Geraldine Carey going to mass.

"I'll walk out with you," he said. "I have a half-day today for myself."

They walked together, to her embarrassment. She glanced at shop-windows to catch a glimpse of their reflection, to see if they looked as awkward as she felt. He was only a head taller than her and part of that was made up by his hard black hat. His clerk's suit was double-breasted, navy-blue with a pale stripe in it, shiny here and there, in need of a good ironing. He wore black leather gloves and carried a walking-stick. He always had the gloves and the walking-stick in church, but his Sunday suit was superior to the one he wore now. Her own fair hair, pinned up under her green-brimmed hat, was what stood out between the two of them. The colour of good corn, Mr. Devereux used to say, and she always considered that a compliment, coming from a grain merchant. Her face was thin and her eyes blue, but reflected in the shop windows there was now only a blur of flesh, a thin shaft between her hat and the green coat that matched it.

"You've had misfortune, Attracta." Solemnly he nodded, repeating the motion of his head until she wished he'd stop. "It was a terrible thing to be killed by mistake."

Attracta didn't know what he was talking about. They passed by the last of the shops in North Street, Shannon's grocery and bar, O'Brien's bakery, the hardware that years ago had run out of stock. The narrow street widened a bit. Mr. Purce said:

"Has she made a Catholic girl of you, Attracta?"

"Who, Mr. Purce?"

"Devereux's woman. Has she tried anything on? Has she shown you rosary beads?"

She shook her head.

"Don't ever look at them if she does. Look away immediately if she gets them out of her apron or anything like that. Will you promise me that, girl?"

"I don't think she would. I don't think Mr. Devereux —"

"You can never tell with that crowd. There isn't a trick in the book they won't hop on to. Will you promise me now? Have nothing to do with carry-on like that."

"Yes, Mr. Purce."

As they walked he prodded at the litter on the pavement with his walking-stick. Cigarette packets and squashed match-boxes flew into the gutter, bits of the *Cork Examiner*, sodden paper-bags. He was known for this activity in the town, and even when he was on his own his voice was often heard protesting at the untidiness.

"I'm surprised they never told you, Attracta," he said. "What are you now, girl?"

"I'm eleven."

"A big girl should know things like that."

"What things, Mr. Purce?"

He nodded in his repetitious manner, and then he explained himself. The tragedy had occurred in darkness, at night: her parents had accidentally become involved with an ambush meant for the Black and Tan soldiers who were in force in the area at the time. She herself had long since been asleep at home, and as he spoke she remembered waking up to find herself in a bed in her aunt's house, without knowing how she got there. "That's how they got killed, Attracta," Mr. Purce said, and then he said an extraordinary thing. "You've got Devereux and his woman to thank for it."

She knew that the Black and Tan soldiers had been camped near the town; she knew there'd been fighting. She realized that the truth about the death had been counted too terrible for a child to bear. But that her parents should have been shot, and shot in error, that the whole thing had somehow been the responsibility of Mr. Devereux and Geraldine Carey, seemed inconceivable to Attracta.

"They destroyed a decent Protestant pair," Mr. Purce continued, still flicking litter from the pavement. "Half-ten at night on a public road, destroyed like pests."

The sun, obscured by clouds while Attracta and Mr. Purce had made the journey from the centre of the town, was suddenly warm on Attracta's face. A woman in a horse and cart, attired in the black hooded cloak of the locality, passed slowly by. There were sacks of meal in the cart which had probably come from Mr. Devereux's mill.

"Do you understand what I'm saying to you, Attracta? Devereux was organizing resistance up in the hills. He had explosives and booby traps, he was drilling men to go and kill people. Did nobody tell you about himself and Geraldine Carey?"

She shook her head. He nodded again, as if to indicate that little better could be expected.

"Listen to me, Attracta. Geraldine Carey was brought into this town by the man she got married to, who used to work at Devereux's mill. Six months later she'd joined up with Devereux in the type of dirty behaviour I wouldn't soil myself telling you about. Not only that, Attracta, she was gun-running with him. She was fixing explosives like a man would, dressed up like a man in uniform. There was nothing Devereux wouldn't do, there was nothing the woman wouldn't do either. They'd put booby traps down and it didn't matter who got killed. They'd ambush the British soldiers when the soldiers didn't have a chance."

It was impossible to believe him. It was impossible to visualize the housekeeper and Mr. Devereux in the role he'd given them. No one with any sense could believe that Geraldine Carey would kill people. Was everything Mr. Purce said a lie? He was a peculiar man: had he some reason for stating her mother and her father had met their deaths in this way?

"Your father was a decent man, Attracta. He was never drunk in his life. There was prayers for him in the chapel, but that was only a hypocrisy of the priests. Wouldn't the priest Quinn like to see every Protestant in this town dead and buried? Wouldn't he like to see you and me six foot down with clay in our eye-sockets?"

Attracta didn't believe that, and more certainly now it seemed to her that everything Mr. Purce said was untrue. Catholics were different; they crossed themselves when they passed their chapel; they went in for crosses and confession; they had masses and candles. But it was hard to accept that Father Quinn, a jovial red-haired man, would prefer it if she were dead. She'd heard her aunt's maid Philomena saying that Father Doran was cantankerous and that Father Martin wasn't worth his salt, but neither of them seemed to Attracta to be the kind of man who'd wish people dead. "Proddy-woddy green guts," Catholic children would shout out sometimes. "Catty-catty going to mass," the Protestants would reply, "riding on the Devil's ass." But there was never much vindictiveness about any of it. The sides were unevenly matched: there were too few Protestants in the town to make a proper opposition; trouble was avoided.

"He was a traitor to his religion, Attracta. And I'll promise you this: if I was to tell you about that woman of his you wouldn't enter the house they have." Abruptly he turned and walked away, back into the town, his walking-stick still frantically working, poking away any litter it could find.

The sun was hot now. Attracta felt sickly within her several layers of clothes. She had a chapter of her history book to read, about the Saxons coming to England. She had four long-division sums to do, and seven lines of poetry to learn. *What potions have I drunk of Syren tears,* the first one stated, a statement Attracta could make neither head nor tail of.

She didn't go straight home. Instead she turned off to the left and walked through a back street, out into the country. She passed fields of mangels and turnips, again trying to imagine the scenes Mr. Purce had sketched for her, the ambush of men waiting for soldiers, the firing of shots. It occurred to her that she had never asked anyone if her parents were buried in the Church of Ireland graveyard.

She passed by tinkers encamped on the verge of the road. A woman ran after her and asked for money, saying her husband had just died. She swore when Attracta said she hadn't any, and then her manner changed again. She developed a whine in her voice, she said she'd pray for Attracta if she'd bring her money, tomorrow or the next day.

Had Mr. Purce only wished to turn her against Mr. Devereux because Mr. Devereux did not go to church? Was there no more to it than that? Did Mr. Purce say the first thing that came into his head?

As Attracta walked, the words of Archdeacon Flower came back to her: in stating that Mr. Devereux was now as gentle as a lamb, was there the implication that once he hadn't been? And had her aunt, worried about Geraldine Carey, been reassured on that score also?

"It's all over now, dear," her aunt said. She looked closely at Attracta and then put her arms round her, as if expecting tears. But tears didn't come, for Attracta was only amazed.

Fifty years later, walking through the heather by the sea, Attracta remembered vividly that moment of her childhood. She couldn't understand how Mr. Devereux and Geraldine Carey had changed so. "Maybe they bear the burden of guilt," Archdeacon Flower had explained, summoned to the house the following day by her aunt. "Maybe they look at you and feel responsible for an accident." What had happened was in the past, he reminded her, as her aunt had. She understood what they were implying, that it must all be forgotten, yet she couldn't help imagining Mr. Devereux and his housekeeper laying booby traps on roads and drilling men in the hills. Geraldine Carey's husband had left the town, Mr. Purce told her on a later occasion: he'd gone to Co. Louth and hadn't been heard of since. "Whore," Mr. Purce said, "No better than a whore she is." Attracta, looking up the word in a dictionary, was astonished.

Having started, Mr. Purce went on and on. Mr. Devereux's house wasn't suitable for an eleven-year old girl to visit, since it was the house of a murderer. Wasn't it a disgrace that a Protestant girl should set foot in a house where the deaths of British soldiers and the Protestant Irish had been planned? One Saturday afternoon, unable to restrain himself, he arrived at the house himself. He shouted at Mr. Devereux from the open hall-door. "Isn't it enough to have destroyed her father and mother without letting that woman steal her for the Pope?" His grey face was suffused beneath his hard hat, his walking-stick thrashed the air. Mr. Devereux called him an Orange Mason. "I hate the bloody sight of you," Mr. Purce said in a quieter voice, and then in his abrupt way he walked off.

That, too, Attracta remembered as she continued her walk around the headland. Mr. Devereux afterwards never referred to it, and Mr. Purce never spoke to her again, as if deciding there was nothing left to say. In the town as she grew up people would reluctantly answer her when she questioned them about her parents' tragedy in an effort to discover more than her aunt or Archdeacon Flower had revealed. But nothing new emerged, the people she asked only agreeing that Mr. Devereux in those days had been as wild as Mr. Purce suggested. He'd drilled the local men, he'd been assisted in every way by Geraldine Carey, whose husband had gone away to Co. Louth. But everything had been different since the night of the tragedy.

Her aunt tried to explain to her the nature of Mr. Purce's hatred of Mr. Devereux. Mr. Purce saw things in a certain light, she said, he could not help himself. He couldn't help believing that Father Quinn would prefer the town's Protestants to be dead and buried. He couldn't help believing that immorality continued in the relationship between Mr. Devereux and his housekeeper when it clearly did not. He found a spark and made a fire of it, he was a bigot and was unable to do anything about it. The Protestants of the town felt ashamed of him.

Mr. Purce died, and was said to have continued in his hatred with his last remaining breaths. He mentioned the Protestant girl, his bleak, harsh voice weakening. She had been contaminated and infected, she was herself no better than the people who used her for their evil purposes. She was not fit to teach the Protestant children of the town, as she was now commencing to do. "As I lie dying," Mr. Purce said to the new clergyman who had succeeded Archdeacon Flower, "I am telling you that, sir." But afterwards, when the story of Mr. Purce's death went round, the people of the town looked at Attracta with a certain admiration, seeming to suggest that for her the twisting of events had not been easy, neither the death of her parents nor the forgiveness asked of her by Mr. Devereux, nor the bigotry of Mr. Purce. She'd been caught in the middle of things, they seemed to suggest, and had survived unharmed.

Surviving, she was happy in the town. Too happy to marry the exchange clerk from the Provincial Bank or the young man who came on a holiday to Cedarstrand with his parents. *Pride goeth before destruction*, her pupils' headlines stated, and *Look before you leap*. Their fingers pressed hard on inky pens, knuckles jutting beneath the strain, tongue-tips aiding concentration. Adriane, Finn Mac Cool, King Arthur's sword, Cathleen ni Houlihan: legends filled the schoolroom, with facts about the Romans and the Normans, square roots and the Gulf Stream. Children grew up and went away, returning sometimes to visit Attracta in her house in North Street. Others remained and in the town she watched them changing, grey coming into their hair, no longer moving as lithely as they had. She developed an affection for the town without knowing why, beyond the fact that it was part of her.

"Yet in all a lifetime I learnt nothing," she said aloud to herself on the headland. "And I taught nothing either." She gazed out at the smooth blue Atlantic but did not see it clearly. She saw instead the brown-paper parcel that contained the biscuit-box she had read about, and the fingers of Penelope Vade undoing the string and brown paper. She saw her lifting off the lid. She saw her frowning for a moment, before the eyes of the man she loved stared deadly into hers. Months later, all courage spent and defeated in her gesture, the body of Penelope Vade dragged itself across the floors of two different rooms. There was the bottle full of aspirins in a cupboard, and water drunk from a Wedgwood-patterned cup, like the cups Attracta drank from every day.

In her schoolroom, with its maps and printed pictures, the sixteen faces stared back at her, the older children at the back. She repeated her question.

"Now, what does anyone think of that?"

Again she read them the news item, reading it slowly because she wanted it to become as rooted in their minds as it was in hers. She lingered over the number of bullets that had been fired into the body of Penelope Vade's husband, and over the removal of his head.

"Can you see that girl? Can you imagine men putting a human head in a tin box and sending it through the post? Can you imagine her receiving it? The severed head of the man she loved?"

"Sure, isn't there stuff like that in the papers the whole time?" one of the children suggested.

She agreed that that was so. "I've had a good life in this town," she added, and the children looked at her as if she'd suddenly turned mad.

"I'm getting out of it," one of them said after a pause. "Back of beyond, miss."

She began at the beginning. She tried to get into the children's minds an image of a baby sleeping while violence and death took place on the Cork road. She described her Aunt Emmeline's house in North Street, the neat feminine house it had been, her aunt's cat Diggory, the small sitting-room, her own very fair hair and her thin face, and the heavy old-fashioned clothes she'd worn in those days. She spoke of the piety of Geraldine Carey, and the grain merchant's tired face. The friendship they offered her was like Penelope Vade proclaiming peace in the city where her husband had been killed; it was a gesture, too.

"His house would smell of roses on a summer's day. She'd carry his meals to him, coming out of the shadows of her kitchen. As if in mourning, the blue blinds darkened the drawing-room. It was they who bore the tragedy, not I."

She described Mr. Purce's face and his grating voice. She tried to make of him a figure they could see among the houses and shops that were familiar to them: the hard black hat, the walking-stick poking away litter. He had done his best to rescue her, acting according to his beliefs. He wanted her not to forget, not realizing that there was nothing for her to remember.

"But I tried to imagine," she said, "as I am asking you to imagine now: my mother and father shot dead on the Cork road, and Mr. Devereux and Geraldine Carey as two monstrous people, and arms being blown off soldiers, and vengeance breeding vengeance."

A child raised a hand and asked to leave the room. Attracta gave permission and awaited the child's return before proceeding. She filled the time in by describing things that had changed in the town, the falling to pieces of O'Mara's Picture House, the closing of the tannery in 1938. When the child came back she told of Mr. Purce's death, how he'd said she was not fit to teach Protestant children.

"I tried to imagine a night I'd heard about," she said, "when Mr. Devereux's men found a man in Madden's public house whom they said had betrayed them, and how they took him out to Cedarstrand and hanged him in a barn. Were they pleased after they'd done that? Did they light cigarettes, saying the man was better off dead? One of those other men must have gone to the post office with the wrapped biscuit-box. He must have watched it being weighed and paid the postage. Did he say to himself he was exceptional to have hoodwinked a post-office clerk?"

Obediently listening in their rows of worn desks, the children wondered what on earth all this was about. No geography or history lesson had ever been so bewildering; those who found arithmetic difficult would have settled for attempting to understand it now. They watched the lined face of their teacher, still thin as she'd said it had been in childhood, the fair hair grey now. The mouth twitched and rapidly moved, seeming sometimes to quiver as if it struggled against tears. What on earth had this person called Penelope Vade to do with anything?

"She died believing that hell had come already. She'd lost all faith in human life, and who can blame her? She might have stayed in Haslemere, like anyone else would have. Was she right to go to the city where her husband had been murdered, to show its other victims that her spirit had not been wholly crushed?"

No one answered, and Attracta was aware of the children's startled gaze. But the startled gaze was a natural reaction. She said to herself that it didn't matter.

"My story is one with hers," she said. "Horror stories, with different endings only. I think of her now and I can see quite clearly the flat she lived in in Belfast. I can see the details, correctly or not I've no idea. Wallpaper with a pattern of brownish purple flowers on it, gaunt furniture casting shadows, a tea-caddy on the hired television set. I drag my body across the floor of two rooms, over a carpet that smells of dust and cigarette ash, over rugs and cool linoleum. I reach up in the kitchen, a hand on the edge of the sink: one by one I eat the aspirins until the bottle's empty."

There was a silence. Feet were shuffled in the schoolroom. No one spoke.

"If only she had known," Attracta said, "that there was still a faith she might have had, that God does not forever withhold His mercy. Will those same men who exacted vengeance on her one day keep bees and budgerigars? Will they serve in shops, and be kind to the blind and the deaf? Will they garden in the evenings and be good fathers? It is not impossible. Oh, can't you see," she cried, "what happened in this town? Here, at the back of beyond. Can't you appreciate it? And can't you see her lying there, mice nibbling her dried blood?"

The children still were quiet, their faces still not registering the comment she wished to make. It was because she'd been clumsy, she thought. All she'd meant to tell them was never to despair. All she meant to do was prepare them for a future that looked grim. She had been happy, she said again. The conversation of Mr. Purce had been full of the truth but it hadn't made any sense because the years had turned the truth around.

To the children she appeared to be talking now to herself. She was old, a few of them silently considered; that was it. She didn't appear to understand that almost every day there was the kind of vengeance she spoke of reported on the television. Bloodshed was wholesale, girls were tarred and left for dead, children no older than them were armed with guns.

Another silence lingered awkwardly and then she nodded at a particular child and the child rose and rang a hand-bell. The children filed away, well-mannered and docile as she had taught them to be. She watched them in the playground, standing in twos and threes, talking about her. It had meant nothing when she'd said that people change. The gleam of hope she'd offered had been too slight to be of use, irrelevant in the horror they took for granted, as part of life. Yet she could not help still believing that it mattered when monsters did not remain monsters for ever. It wasn't much to put against the last bleak moments of Penelope Vade, but it was something for all that. She wished she could have made her point.

Twenty minutes later, when the children returned to the schoolroom, her voice no longer quivered, nor did it seem to struggle against tears. The older children learnt about agriculture in Sweden, the younger ones about the Pyrenees, the youngest that Munster had six countries. The day came to an end at three o'clock and when all the children had gone Attracta locked the schoolroom and walked to the house she had inherited in North Street.

A week later Archdeacon Flower's successor came to see her, his visit interrupting further violence on the television news. He beat about the bush while he nibbled biscuits and drank cups of tea by the fire; then suggested that perhaps she should consider retiring one of these days. She was over sixty, he pointed out with his clerical laugh, and she replied that Mr. Ayrie had gone on until seventy. Sixty, the clergyman repeated with another laugh, was the post's retirement age. Children were a handful nowadays.

She smiled, thinking of her sixteen docile charges. They had chattered to their parents, and the parents had been shocked to hear that they'd been told of a man decapitated and a girl raped seven times. School was not for that, they had angrily protested to the clergyman, and he had had no option but to agree. At the end of the summer term there'd be a presentation of Waterford glass.

"Every day in my schoolroom I should have honoured the small, remarkable thing that happened in this town. It matters that she died in despair, with no faith left in human life."

He was brisk. For as long most people could remember she had been a remarkable teacher; in no way had she failed. He turned the conversation to more cheerful topics, he ate more biscuits and a slice of cake. He laughed and even made a joke. He retailed a little harmless gossip.

Eventually she stood up. She walked with her visitor to the hall, shook hands with him and saw him out. In the sitting-room she piled the tea things on to a tray and placed it on a table by the door. She turned the television on again but when the screen lit up she didn't notice it. The face of Penelope Vade came into her mind, the smile a little crooked, the freckled cheeks.

Social
COMMITMENT
and the Latin American Writer

MARIO VARGAS LLOSA

Mario Vargas Llosa was born in Arequipa, Peru in 1936. Of the politics of imaginative writing he writes: "By spurring the imagination, fiction both assuages human dissatisfaction and simultaneously incites it... One can well understand why regimes that seek to exercise total control over life mistrust works of fiction and subject them to censorship. Emerging from one's self, being another, even in illusion, is a way of being less a slave and of experiencing the risks of freedom."

THE PERUVIAN NOVELIST JOSE MARIA Arguedas killed himself on the second day of December 1969 in a classroom of La Molina Agricultural University in Lima. He was a very discreet man, and so as not to disturb his colleagues and the students with his suicide, he waited until everybody had left the place. Near his body was found a letter with very detailed instructions about his burial — where he should be mourned, who should pronounce the eulogies in the cemetery — and he asked too that an Indian musician friend of his play the *huaynos* and *mulizas* he was fond of. His will was respected, and Arguedas, who had been, when he was alive, a very modest and shy man, had a very spectacular burial.

But some days later other letters written by him appeared, little by little. They too were different aspects of his last will, and they were addressed to very different people: his publisher, friends, journalists, academics, politicians. The main subject of these letters was his death, of course, or better, the reasons for which he decided to kill himself. These reasons changed from letter to letter. In one of them he said that he had decided to commit suicide because he felt that he was finished as a writer, that he no longer had the impulse and the will to create. In another he gave moral, social and political reasons: he could no longer stand the misery and neglect of the Peruvian peasants, those people of the Indian communities among whom he had been raised; he lived oppressed and anguished by the crises of the cultural and educational life in the country; the low level and abject nature of the press and the caricature of liberty in Peru were too much for him, et cetera.

In these dramatic letters we follow, naturally, the personal crises that Arguedas had been going through, and they are the desperate call of a suffering man who, at the edge of the abyss, asks mankind for help and compassion. But they are not only that: a clinical testimony. At the same time, they are graphic evidence of the situation of the writer in Latin America, of the difficulties and pressures of all sorts that have surrounded and oriented and many times destroyed the literary vocation in our countries.

In the USA, in Western Europe, to be a writer means, generally, first (and usually only) to assume a personal responsibility. That is, the responsibility to achieve in the most rigorous and authentic way a work which, for its artistic values and originality, enriches the language and culture of one's country. In Peru, in Bolivia, in Nicaragua, et cetera, on the contrary, to be a writer means, at the same time, to assume a social responsibility; at the same time that you develop a personal literary work, you should serve, through your writing but also through your actions, as an active participant in the solution of the economic, political and cultural problems of your society. There is no way to escape this obligation. If you tried to do so, if you were to isolate yourself and concentrate exclusively on your own work, you would be severely censured and considered, in the best of cases, irresponsible and selfish, or at worst, even by omission, an accomplice to all the evils — illiteracy, misery, exploitation, injustice, prejudice — of your country and against which you have refused to fight. In the letters which he wrote once he had prepared the gun with which he was to kill himself, Arguedas was trying, in the last moments of his life, to fulfill this moral imposition that impels all Latin American writers to social and political commitment.

Why is it like this? Why cannot writers in Latin America, like their American and European colleagues, be artists, and only artists? Why must they also be reformers, politicians, revolutionaries, moralists? The answer lies in the social conditions of Latin America, the problems which face our countries. All countries have problems, of course, but in many parts of Latin America, both in the past and in the present, the problems which constitute the closest daily reality for people are not freely discussed and analysed in public, but are usually denied and silenced. There are no means through which those problems can be presented and denounced, because the social and political establishment exercises a strict censorship of the media and over all the communications systems. For example, if today you hear Chilean broadcasts or see Argentine television, you won't hear a word about the political prisoners, about the exiles, about the torture, about the violations of human rights in those two countries that have outraged the conscience of the world. You will, however, be carefully informed, of course, about the iniquities of the communist countries. If you read the daily newspapers of my country, for instance — which have been confiscated by the government, which now controls them — you will not find a word about the continuous arrests of labour leaders or about the murderous inflation that affects everyone. You will read only about what a happy and prosperous country Peru is and how much we Peruvians love our military rulers.

What happens with the press, TV and radio happens too, most of the time, with the universities. The government persistently interferes with them; teachers and students considered subversive or hostile to the official system are expelled and the whole curriculum reorganized according to political considerations. As an indication of what extremes of absurdity this "cultural policy" can reach, you must remember, for instance, that in Argentina, in Chile and in Uruguay the departments of Sociology have been closed indefinitely, because the social sciences are considered subversive. Well, if academic institutions submit to this manipulation and censorship, it is improbable that contemporary political, social and economic problems of the country can be described and discussed freely. Academic knowledge in many Latin American countries is, like the press and the media, a victim of the deliberate turning away from what is actually happening in society. This vacuum has been filled by literature.

This is not a recent phenomenon. Even during the Colonial Period, though more especially since Independence (in which intellectuals and writers played an important role), all over Latin America novels, poems and plays were — as Stendhal once said he wanted the novel to be — the mirrors in which Latin Americans could truly see their faces and examine their sufferings. What was, for political reasons, repressed or distorted in the press and in the schools and universities, all the evils that were buried by the military and economic elite which ruled the countries, the evils which were never mentioned in the speeches of the politicians nor taught in the lecture halls nor criticised in the congresses nor discussed in magazines found vehicle of expression in literature.

So, something curious and paradoxical occurred. The realm of imagination became in Latin America the kingdom of objective reality; fiction became a substitute for social science; our best teachers about reality were dreamers, the literary artists. And this is true not only for our great essayists — such as Sarmiento, Martí, González Prada, Rodó, Vasconcelos, José Carlos Mariátegui — whose books are indispensable for a thorough comprehension of the historical and social reality of their respective countries, but it is also valid for the writers who only practised the creative literary genres: fiction, poetry and drama. We can say without exaggeration that the most representative and genuine description of the real problems of Latin America during the nineteenth century is to be found in literature, and that it was in the verses of the poets or the plots of the novelists that, for the first time, the social evils of Latin America were denounced.

We have a very illustrative case with what is called *indigenismo*, the literary current which, from the middle of the nineteenth century until the first decades of our century focused on the Indian peasant of the Andes and his problems as its main subject. The indigenist writers were the first people in Latin America to describe the terrible conditions in which the Indians were still living three centuries after the Spanish conquest, the impunity with which they were abused and exploited by the landed proprietors — the *latifundistas,* the *gamonales* — men who sometimes owned land areas as big as a European country, where they were absolute kings who treated their Indians worse and sold them cheaper than their cattle. The first indigenist writer was a woman, an energetic and enthusiastic reader of the French novelist Émile Zola and the positivist philosophers: Clorinda Matto de Turner (1854-1909). Her novel *Aves sin nido* opened a road of social commitment to the problems and aspects of Indian life that Latin American writers would follow, examining in detail and from all angles, denouncing injustices and praising and rediscovering the values and traditions of an Indian culture which until then, at once incredibly and ominously, had been systematically ignored by the official culture. There is no way to research and analyse the rural history of the continent and to understand the tragic destiny of the inhabitants of the Andes since the regions ceased to be a colony without going through their (the indigenists') books. These constitute the best — and sometimes the only — testimony to this aspect of our reality.

Am I saying, then, that because of the authors' moral and social commitment this literature is good literature? That because of their generous and courageous goals of breaking the silence about the real problems of society and of contributing to the solution of these problems, this literature was an artistic accomplishment? Not at all. What actually happened in many cases was the contrary. The pessimistic dictum of André Gide, who once said that with good sentiments one has bad literature, can be, alas, true. Indigenist literature is very important from a historical and social point of view, but only in exceptional cases is it of literary importance. These novels or poems written, in general, very quickly, impelled by the present situation, with militant passion, obsessed with the idea of denouncing a social evil, of correcting a wrong, lack most of what is essential in a work of art: richness of expression, technical originality. Because of the didactic intentions they become simplistic and superficial; because of their political partisanship they are sometimes demagogic and melodramatic; and because of their nationalist or regionalist scope they can be very provincial and quaint. We can say that many of these writers, in order to serve better moral and social needs, sacrificed their vocation on the altar of politics. Instead of artists, they chose to be moralists, reformers, politicians, revolutionaries.

You can judge from your own particular system of values whether this sacrifice is right or wrong, whether the immolation of art for social and political aims is worthwhile or not. I am not dealing at the moment with this problem. What I am trying to show is how the particular circumstances of Latin American life have traditionally oriented literature in this direction and how this has created for writers a very special situation. In one sense people — the real or potential readers of the writer — are accustomed to considering literature as something intimately associated with living and social problems, the activity through which all that is repressed or disfigured in society will be named, described and condemned. They expect novels, poems and plays to counterbalance the policy of disguising and deforming reality which is current in the official culture and to keep alive the hope and spirit of change and revolt among the victims of that policy. In another sense this confers on the writer, as a citizen, a kind of moral and spiritual leadership, and he must try, during his life as a writer, to act according to this image of the role he is expected to play. Of course he can reject it and refuse this task that society wants to impose on him; and declaring that he does not want to be either a politician or a moralist or a sociologist, but only an artist, he can seclude himself in his personal dreams. However, this will be considered (and in a way, it is) a political, a moral and a social choice. He will be considered by his real and potential readers a deserter and a traitor, and his poems, novels and plays will be endangered. To be an artist, only an artist, can become, in our countries, a kind of moral crime, a political sin. All our literature is marked by this fact, and if this is not taken into consideration, one cannot fully understand all the differences that exist between it and other literatures in the world.

No writer in Latin America is unaware of the pressure that is put on him pushing him to a social commitment. Some accept this because the external impulse coincides with their innermost feelings and personal convictions. These cases are, surely, the happy ones. The coincidence between the individual choice of the writer and the idea that society has of his vocation permits the novelist, poet or playwright to create freely, without any pangs of conscience, knowing that he is supported and approved by his contemporaries. It is interesting to note that many Latin American men and women whose writing started out as totally uncommitted, indifferent or even hostile to social problems and politics, later — sometimes gradually, sometimes abruptly — oriented their writings in this direction. The reason for this change could be, of course, that they adopted new attitudes, acknowledging the terrible social problems of our countries, an intellectual discovery of the evils of society and the moral decision to fight them. But we cannot dismiss the possibility that in this change (conscious or unconscious) the psychological and practical trouble it means for a writer to resist the social pressure for political commitment also played a role, as did the psychological and practical advantages which led him to act and to write as society expects him to.

All this has given Latin American literature peculiar features. Social and political problems constitute a central subject for it, and they are present everywhere, even in works where, because of the theme and form, one would never expect to find them. Take the case, for example, of the "literature of fantasy" as opposed to "realist literature." This kind of literature, whose raw material is subjective fantasy, does not reflect, usually, the mechanisms of economic injustice in society nor the problems faced by urban and rural workers which make up the objective facts of reality; instead — as in Edgar Allan Poe or Villiers de L'Isle-Adam — this literature builds a new reality, essentially different from "objective reality," out of the most intimate obsessions of writers. But in Latin America (mostly in modern times, but also in the past) fantastic literature also has its roots in objective reality and is a vehicle for exposing social and political evils. So, fantastic literature becomes, in this way, symbolical literature in which, disguised with the prestigious clothes of dreams and unreal beings and facts, we recognize the characters and problems of contemporary life.

We have many examples among contemporary Latin American writers of this "realistic" utilization of unreality. The Venezuelan Salvador Garmendia has described, in short stories and novels of nightmarish obsessions and impossible deeds, the cruelty and violence of the streets of Caracas and the frustrations and sordid myths of the lower middle classes of that city. In the only novel of the Mexican Juan Rulfo, *Pedro Páramo* (1955) — all of whose characters, the reader discovers in the middle of the book, are dead people — fantasy and magic are not procedures to escape social reality; on the contrary, they are simply alternative means to represent the poverty and sadness of life for the peasants of a small Jalisco village.

Another interesting case is Julio Cortázar. In his first novels and short stories we enter a *fantastic* world, which is very mischievous because it is ontologically different from the world that we know by reason and experience yet has, at first approach, all the appearances —

features — of real life. Anyway, in this world social problems and political statements do not exist; they are aspects of human experience that are omitted. But in his more recent books — and principally in the latest novel, *Libro de Manuel* (1973) — politics and social problems occupy a place as important as that of pure fantasy. The "fantastic element" is merged, in this novel, with statements and motifs which deal with underground militancy, terrorism, revolution and dictatorship.

What happens with prose also happens with poetry, and as among novelists, one finds this necessity for social commitment in all kinds of poets, even in those whom, because of the nature of their themes, one would expect not to be excessively concerned with militancy. This is what occurred, for instance, with religious poetry, which is, in general, very publicized in Latin America. And it is symptomatic that, since the death of Pablo Neruda, the most widely known poet — because of his political radicalism, his revolutionary lyricism, his colourful and schematic ideology — is a Nicaraguan priest, a former member of the American Trappist monastery of Gethsemane: Ernesto Cardenal.

It is worth noting too that the political commitment of writers and literature in Latin America is a result not only of the social abuse and economic exploitation of large sectors of the population by small minorities and brutal military dictatorships. There are also cultural reasons for this commitment, exigencies that the writer himself sees grow and take root in his conscious during and because of his artistic development. To be a writer, to discover this vocation and to choose to practise it pushes one inevitably, in our countries, to discover all the handicaps and miseries of underdevelopment. Inequities, injustice, exploitation, discrimination, abuse are not only the burden of peasants, workers, employees, minorities. They are also social obstacles for the development of a cultural life. How can literature exist in a society where the rates of illiteracy reach fifty or sixty percent of the population? How can literature exist in countries where there are no publishing houses, where there are no literary publications, where if you want to publish a book you must finance it yourself? How can a cultural and literary life develop in a society where the material conditions of life — lack of education, subsistence wages, et cetera —

establish a kind of cultural apartheid, that is, prevent the majority of the inhabitants from buying and reading books? And if, besides all that the political authorities have established a rigid censorship in the press, in the media and in the universities, that is, in those places through which literature would normally find encouragement and an audience, how could the Latin American writer remain indifferent to social and political problems? In the practice itself of his art — in the obstacles that he finds for this practice — the Latin American writer finds reasons to become politically conscious and to submit to the pressures of social commitment.

We can say that there are some positive aspects in this kind of situation for literature. Because of that commitment, literature is forced to keep in touch with living reality, with the experiences of people, and it is prevented from becoming — as unfortunately has happened in some developed societies — an esoteric and ritualistic experimentation in new forms of expression almost entirely dissociated from real experience. And because of social commitment, writers are obliged to be socially responsible for what they write and for what they do, because social pressure provides a firm barrier against the temptation of using words and imagination in order to play the game of moral irresponsibility, the game of the *enfant terrible* who (only at the level of words, of course) cheats, lies, exaggerates and proposes the worst options.

But this situation has many dangers, too. The function and the practice of literature can be entirely distorted if the creative writings are seen only (or even mainly) as the materialization of social and political aims. What is to be, then, the borderline, the frontier between history, sociology and literature? Are we going to say that literature is only a degraded form (since its data are always dubious because of the place that fantasy occupies in it) of the social sciences? In fact, this is what literature becomes if its most praised value is considered to be the testimony it offers of objective reality, if it is judged principally as a true record of what happens in society.

On the other hand, this opens the door of literature to all kinds of opportunistic attitudes and intellectual blackmail. How can I condemn as an artistic failure a novel that explicitly protests against *ríos profundos*, [1958]) a melancholic escape to the days and places of his childhood, the world of the little Indian villages — San Juan de

Lucanas, Puquio — or towns of the Andes such as Abancay, whose landscapes and customs he described in a tender and poetic prose. But later he felt obliged to renounce this kind of lyric image to fill the social responsibilities that everybody expected of him. And he wrote a very ambitious book, *Todas las sangres* (1964), in which he tried, escaping from himself, to describe the social and political problems of his country. The novel is a total failure: the vision is simplistic and even a caricature. We find none of the great literary virtues that made his previous books genuine works of art. The book is the classic failure of an artistic talent due to the self-imposition of social commitment. The other books of Arguedas oscillate between those two sides of his personality, and it is probable that all this played a part in his suicide.

When he pressed the trigger of the gun, at the University of La Molina, on the second day of December in 1969, José Mariá Arguedas was too, in a way, showing how difficult and daring it can be to be a writer in Latin America.

ORION

JEANETTE WINTERSON

Jeanette Winterson was born in Lancashire, England in 1959. For the writer in an invented country, she says "the journey itself is that language is just ahead and a bit out of our reach." To the reader, her voice in her novel, The Passion, simply entreats: "I'm telling you stories. Trust me."

HERE ARE THE COORDINATES: FIVE HOURS, thirty minutes right ascension (the coordinate on the celestial sphere analogous to longitude on earth) and zero declination (at the celestial equator). Any astronomer can tell where you are.

It's different isn't it from head back in the garden on a frosty night sensing other worlds through a pair of binoculars? I like those nights. Kitchen light out and wearing Wellingtons with shiny silver insoles. On the wrapper there's an astronaut showing off his shiny silver suit. A short trip to the moon has brought some comfort back to earth. We can wear what Neil Armstrong wore and never feel the cold. This must be good news for star-gazers whose feet are firmly on the ground. We have moved with the times. And so will Orion.

Every 200 000 years or so, the individual stars within each constellation shift position. That is, they are shifting all the time, but more subtly than any tracker dog of ours can follow. One day, if the earth has not voluntarily opted out of the solar system, we will wake up to a new heaven whose dome will again confound us. It will still be home but not a place to take for granted. I wouldn't be able to tell you the story of Orion and say, "Look, there he is, and there's his dog Sirius whose loyalty has left him bright." The dot-to-dot logbook of who we were is not a fixed text. For Orion, who was the result of three of the gods in a good mood pissing on an ox-hide, the only tense he recognized was the future continuous. He was a mighty hunter. His arrow was always in flight, his prey endlessly just ahead of him. The carcasses he left behind became part of his past faster than they could decay. When he went to Crete he didn't do any sunbathing. He rid the island of all its beasts. He could really swing a cudgel.

Stories abound. Orion was so tall that he could walk along the sea bed without wetting his hair. So strong he could part a mountain. He wasn't the kind of man who settles down. And then he met Artemis, who wasn't the kind of woman who settles down either. They were both hunters and both gods. Their meeting is recorded in the heavens, but you can't see it every night, only on certain nights of the year. The rest of the time Orion does his best to dominate the skyline, as he always did.

Our story is the old clash between history and home. Or to put it another way, the immeasurable impossible space that seems to divide the hearth from the quest.

Listen to this.

On a wild night, driven more by weariness than by common sense, King Zeus decided to let his daughter do it differently: she didn't want to get married and sit out some war while her man, god or not, underwent the ritual metamorphosis from palace prince to craggy hero: she didn't want her children. She wanted to hunt. Hunting did her good.

By morning she had packed and set off for a new life in the woods. Soon her fame spread and other women joined her but Artemis didn't care for company. She wanted to be alone. In her solitude she discovered something very odd. She had envied men their long-legged freedom to roam the world and return full of glory to wives who only waited. Without rejecting it, she had simply hoped to take on the freedoms that belonged to the other side. What if she travelled the world and the seven seas like a hero? Would she find something different or the old things in different disguises? She found that the whole world could be contained within one place because that place was herself. Nothing had prepared her for this.

The alchemists have a saying, *Tertium non data*. The third is not given. That is, the transformation from one element into another, from waste matter into best gold is a process that cannot be documented. It is fully mysterious. No one really knows what effects the change. And so it is with the mind that moves from its prison to a vast plain without any movement at all. We can only guess at what happened.

One evening when Artemis had lost her quarry, she lit a fire where she was and tried to rest. But the night was shadowy and full of games. She saw herself by the fire: as a child, a woman, a hunter, a queen. Grabbing the child, she lost sight of the woman and when she drew her bow the queen fled. What would it matter if she crossed the world and hunted down every living creature as long as her separate selves eluded her? In the end when no one was left, she would have to confront herself. Leaving home meant leaving nothing behind. It came too, all of it, and waited in the dark. She realized that the only war worth fighting was the one that raged within; the rest were all diversions. In this small place, her hunting miles, she was going to bring herself home. Home was not a place for the faint-hearted; only the very brave could live with themselves.

In the morning she set out, and set out every morning day after day.

In her restlessness she found peace.

Then Orion came.

He wandered into Artemis's camp, scattering her dogs and bellowing like a bad actor, his right eye patched and his left arm in a splint. She was a mile or so away fetching water. When she returned she saw this huge rag of a man eating her goat. Raw. When he'd finished with a great belch and the fat still fresh around his mouth, he suggested they take a short stroll by the sea's edge. Artemis didn't want to but she was frightened. His reputation hung around him like bad breath. The ragged shore, rock-pitted and dark with weed, reminded him of his adventures and he unravelled them in detail while the tide came in up to her waist. There was nowhere he hadn't been, nothing he hadn't seen. He was faster than a hare and stronger than a pair of bulls. He was as good as a god.

"You smell," said Artemis, but he didn't hear.

Eventually he allowed her to wade in from the rising water and light a fire for them both. No, he didn't want her to talk; he knew about her already. He'd been looking for her. She was a curiosity; he was famous. What a marriage.

But Artemis did talk. She talked about the land she loved and its daily changes. This was where she wanted to stay until she was ready to go. The journey itself was not enough. She spoke quickly, her words hanging on to each other; she'd never told anyone before. As she said it, she knew it was true, and it gave her strength to get up and say goodbye. She turned. Orion raped Artemis and fell asleep.

She thought about that time for years. It took a few moments only and she was really aware of only the hair of his stomach that was matted with sand, scratching her skin. When he'd finished, she pushed him off already snoring. His snores shook the earth. Later, in the future, the time would remain vivid and unchanged. She wouldn't think of it differently; she wouldn't make it softer or harder. She would just keep it and turn it over in her hands. Her revenge had been swift, simple and devastatingly ignominious. She killed him with a scorpion.

In a night, 200 000 years can pass, time moving only in our minds. The steady marking of the seasons, the land well-loved and always changing, continues outside, while inside, light years move us on to landscapes that revolve under different skies.

Artemis lying beside dead Orion sees her past changed by a single act. The future is still intact, still unredeemed, but the past is irredeemable. She is not who she thought she was. Every action and decision has led her here. The moment has been waiting the way the top step of the stairs waits for the sleepwalker. She had fallen and now she is awake. As she looks at the sky, the sky is peaceful and exciting. A black cloak pinned with silver brooches that never need polish. Somebody lives there for sure, wrapped up in the glittering folds. Somebody who recognized that the journey by itself is never enough and gave us spaceships long ago in favour of home.

On the beach the waves made pools around Artemis's feet. She kept the fire burning, warming herself and feeling Orion growing slowly cold. It takes time for the body to stop playing house.

The fiery circle that surrounded her contained all the clues to recognize that life is for a moment in one shape then released into another. Monuments and cities would fade away like the people who built them. No resting place or palace could survive the light years ahead. There was no history that would not be rewritten, and the earliest days were already too far away to see. What would history make of tonight?

Tonight is clear and bright with a cold wind stirring the waves into peaks. The foam leaves slug trails in rough triangles on the sand. The salt smell bristles the hair inside her nostrils; her lips are dry. She's thinking about her dogs. They feel like home because she feels like home. The stars show her how to hang in space supported by nothing at all. Without medals or certificates or territories she owns, she can burn as they do, travelling through time until time stops and eternity changes things again. She hasn't noticed that change doesn't hurt her.

It's almost light, which means the disappearing act will soon begin. She wants to lie awake watching until the night fades and the stars fade and first grey-blue slates the sky. She wants to see the sun slash the water. But she can't stay awake for everything; some things have to pass her by. So what she doesn't see are the lizards coming out for food or Orion's eyes turned glassy overnight. A small bird perches on his shoulder, trying to steal a piece of his famous hair.

Artemis waited until the sun was up before she trampled out the fire. She brought rocks and stones to cover Orion's body from the eagles. She made a high mound that broke the thudding wind as it scored the shore. It was a stormy day, black clouds and a thick orange glow on the horizon. By the time she had finished, she was soaked with rain. Her hands were bleeding, her hair kept catching in her mouth. She was hungry but not angry now.

The sand that had been blonde yesterday was now brown with wet. As far as she could see there was the grey water white-edged and the birds of prey wheeling above it. Lonely cries and she was lonely, not for friends but for a time that hadn't been violated. The sea was hypnotic. Not the wind or the cold could move her from where she sat like one who waited. She was not waiting; she was remembering. She was trying to find what it was that had brought her here. The third is not given. All she knew was that she had arrived at the frontiers of common sense and crossed over. She was safe now. No safety without risk, and what you risk reveals what you value.

She stood up and in the getting-dark walked away, not looking behind her but conscious of her feet shaping themselves in the sand. Finally, at the headland, after a bitter climb to where the woods bordered the steep edge, she turned and stared out, seeing the shape of Orion's mound, just visible now, and her own footsteps walking away. Then it was fully night, and she could see nothing to remind her of the night before except the stars.

And what of Orion? Dead but not forgotten. For a while he was forced to pass the time in Hades, where he beat up flimsy beasts and cried a lot. Then the gods took pity on him and drew him up to themselves and placed him in the heavens for all to see. When he rises at dawn, summer is nearly here. When he rises in the evening, beware of winter and storms. If you see him at midnight, it's time to pick the grapes. He has his dogs with him, Canis Major and Canis Minor and Sirius, the brightest star in our galaxy. Under his feet, if you care to look, you can see a tiny group of stars: Lepus, the hare, his favourite food.

Orion isn't always home. Dazzling as he is, like some fighter pilot riding the sky, he glows very faint, if at all, in November. November being the month of Scorpio.

How
WANG FO
Was Saved

MARGUERITE YOURCENAR

Marguerite Yourcenar was born in Brussels, Belgium in 1903. She was a novelist, critic, biographer, translator, interpreter, poet, and playwright and was the first woman ever to be elected to the Academie Francaise in 1981. She identifies this tale as a transcription of a Taoists fable of ancient China,"more or less freely developed by myself;" it is "a reminder of how strange each existence is, where everything floats past like an ever-flowing stream and only those things which matter, instead of sinking to the depths, rise to the surface and finally reach, together with us, the sea" Yourcenar died near her island-home in Maine in 1987.

THE OLD PAINTER WANG-FO AND HIS disciple Ling were wandering along the roads of the Kingdom of Han.

They made slow progress because Wang-Fo would stop at night to watch the stars and during the day to observe the dragonflies. They carried hardly any luggage, because Wang-Fo loved the image of things and not the things themselves, and no object in the world seemed to him worth buying, except brushes, pots of lacquer and China ink, and rolls of silk and rice paper. They were poor, because Wang-Fo would exchange his paintings for a ration of boiled millet, and paid no attention to pieces of silver. Ling, his disciple, bent beneath the weight of a sack full of sketches, bowed his back with respect as if he were carrying the heavens' vault, because for Ling the sack was full of snow-covered mountains, torrents in spring, and the face of the summer moon.

Ling had not been born to trod down the roads, following an old man who seized the dawn and captured the dusk. His father had been a banker who dealt in gold, his mother the only child of a jade merchant who had left her all his worldly possessions, cursing her for not being a son. Ling had grown up in a house where wealth made him shy: he was afraid of insects, of thunder and the face of the dead. When Ling was fifteen, his father chose a bride for him, a very beautiful one because the thought of the happiness he was giving his son consoled him for having reached the age in which the night is meant for sleep. Ling's wife was as frail as a reed, childish as milk, sweet as saliva, salty as tears. After the wedding, Ling's parents became discreet to the point of dying, and their son was left alone in a house painted vermilion, in the company of his young wife who never stopped smiling and a plum tree that blossomed every spring with pale-pink flowers. Ling loved this woman of crystal-clear heart as one loves a mirror that will never tarnish, or a talisman that will protect one forever. He visited the teahouses to follow the debates of fashion, and only moderately favoured acrobats and dangers.

One night, in the tavern, Wang-Fo shared Ling's table. The old man had been drinking in order to better paint a drunkard, and he cocked his head to one side as if trying to measure the distance between his hand and his bowl. The rice wine undid the tongue of the taciturn craftsman, and that night Wang-Fo spoke as if silence were a wall and words the colours with which to cover it. Thanks to him, Ling got to know the beauty of the drunkards' faces blurred by the vapours of hot drink, the brown splendour of the roasts unevenly brushed by tongues of fire, and the exquisite blush of wine stains strewn on the tablecloths like withered petals. A gust of wind broke the window: the downpour entered the room. Wang-Fo leaned out to make Ling admire the livid zebra stripes of lightning, and Ling, spellbound, stopped being afraid of storms.

Ling paid the old painter's bill, and as Wang-Fo was both without money and without lodging, he humbly offered him a resting place. They walked away together; Ling held a lamp whose light projected unexpected fires in the puddles. That evening, Ling discovered with surprise that the walls of his house were not red, as he had always thought, but the colour of an almost rotten orange. In the courtyard, Wang-Fo noticed the delicate shape of a bush to which no one had paid any attention until then, and compared it to a young woman letting down her hair to dry. In the passageway, he followed with delight the hesitant trail of an ant along the cracks in the wall, and Ling's horror of these creatures vanished into thin air. Realizing that Wang-Fo had just presented him with the gift of a new soul and a new vision of the world, Ling respectfully offered the old man the room in which his father and mother had died.

For many years now, Wang-Fo had dreamed of painting the

portrait of a princess of olden days playing the lute under a willow. No woman was sufficiently unreal to be his model, but Ling would do because he was not a woman. Then Wang-Fo spoke of painting a young prince shooting an arrow at the foot of a large cedar tree. No young man of the present was sufficiently unreal to serve as his model, but Ling got his own wife to pose under the plum tree in the garden. Later on, Wang-Fo painted her in a fairy costume against the clouds of twilight, and the young women wept because it was an omen of death. As Ling came to prefer the portraits painted by Wang-Fo to the young woman herself, her face began to fade, like a flower exposed to warm winds and summer rains. One morning, they found her hanging from the branches of the pink plum tree: the ends of the scarf that was strangling her floated in the wind, entangled with her hair. She looked even more delicate than usual, and as pure as the beauties celebrated by the poets of days gone by. Wang-Fo painted her one last time, because he loved the green hue that suffuses the face of the dead. His disciple Ling mixed the colours and the task needed such concentration that he forgot to shed tears.

One after the other, Ling sold his slaves, his jades, and the fish in his pond to buy his master pots of purple paint that came from the West. When the house emptied, they left it, and Ling closed the door of his past behind him. Wang-Fo felt weary of a city where the faces could no longer teach him secrets of ugliness or beauty, and the master and his disciple walked away together down the roads of the Kingdom of Han.

Their reputation preceded them into the villages, to the gateway of fortresses, and into the atrium of temples where restless pilgrims halt at dusk. It was murmured that Wang-Fo had the power to bring his paintings to life by adding a last touch of colour to their eyes. Farmers would come and beg him to paint a watchdog, and lords would ask him for portraits of their best warriors. The priests honoured Wang-Fo as a sage; the people feared him as a sorcerer. Wang-Fo enjoyed these differences of opinion which gave him the chance to study expressions of gratitude, fear, and veneration.

Ling begged for food, watched over his master's rest, and took advantage of the old man's raptures to massage his feet. With the first rays of the sun, when the old man was still asleep, Ling went in pursuit of timid landscapes hidden behind bunches of reeds. In the evening when the master, disheartened, threw down his brushes, he would carefully pick them up. When Wang-Fo became sad and spoke of his old age, Ling would smile and show him the solid trunk of an old oak; when Wang-Fo felt happy and made jokes, Ling would humbly pretend to listen.

One day, at sunset, they reached the outskirts of the Imperial City and Ling sought out and found an inn in which Wang-Fo could spend the night. The old man wrapped himself up in rags, and Ling lay down next to him to keep him warm because spring had only just begun and the floor of beaten earth was still frozen. At dawn, heavy steps echoed in the corridors of the inn; they heard the frightened whispers of the innkeeper and orders shouted in a foreign, barbaric tongue. Ling trembled, remembering that the night before he had stolen a rice cake for his master's supper. Certain that they would come to take him to prison, he asked himself who would help Wang-Fo ford the next river on the following day.

The soldiers entered carrying lanterns. The flames gleaming through the motley paper cast red and blue lights on their leather helmets. The string of a bow quivered over their shoulders, and the fiercest among them suddenly let out a roar for no reason at all. A heavy hand fell on Wang-Fo's neck, and the painter could not help noticing that the soldiers' sleeves did not match the colour of their coats.

Helped by his disciple, Wang-Fo followed the soldiers, stumbling along uneven roads. The passing crowds made fun of these two criminals who were certainly going to be beheaded. The soldiers answered Wang-Fo's questions with savage scowls. His bound hand hurt him, and Ling in despair looked smiling at his master, which for him was a gentler way of crying.

They reached the threshold of the Imperial Palace, whose purple walls rose in broad daylight like a sweep of sunset. The soldiers led Wang-Fo through countless square and circular rooms whose shapes symbolized the seasons, the cardinal points, the male and the female, longevity, and the prerogatives of power. The doors swung on their hinges with a musical note, and were placed in such a manner that one followed the entire scale when crossing the palace from east to west. Everything combined to give an impression of superhuman power and subtlety, and one could feel that here the simplest orders were as final and as terrible as the wisdom of the ancients. At last, the air became thin and the silence so deep that not even a man under torture would have dared to scream. A eunuch lifted a tapestry; the soldiers began to tremble like women, and the small troop entered the chamber in which the Son of Heaven sat on a high throne.

It was a room without walls, held up by thick columns of blue stone. A garden spread out on the far side of the marble shafts, and each and every flower blooming in the greenery belonged to a rare species brought here from across the oceans. But none of them had any perfume, so that the Celestial Dragon's meditations would not be troubled by fine smells. Out of respect for the silence in which his thoughts evolved, no bird had been allowed within the enclosure, and even the the bees had been driven away. An enormous wall separated the garden from the rest of the world, so that the wind that sweeps over dead dogs and corpses on the battlefield would not dare brush the Emperor's sleeve.

The Celestial Master sat on a throne of jade, and his hands were wrinkled like those of an old man, though he had scarcely reached the age of twenty. His robe was blue to symbolize winter, and green to remind one of spring. His face was beautiful but blank, like a looking glass placed too high, reflecting nothing except the stars and the immutable heavens. To his right stood his Minister of Perfect Pleasures, and to his left his Counsellor of Just Torments. Because his courtiers, lined along the base of the column, always lent a keen ear to the slightest sound from his lips, he had adopted the habit of speaking in a low voice.

"Celestial Dragon," said Wang-Fo, bowing low, "I am old, I am poor, I am weak. You are like summer; I am like winter. You have Ten Thousand Lives; I have but one, and it is near its close. What have I done to you? My hands have been tied, these hands that never harmed you."

"You ask what you have done to me, old Wang-Fo?" said the Emperor.

His voice was so melodious that it made one want to cry. He raised his right hand, to which the reflections from the jade pavement gave a pale sea-green hue like that of an underwater plant, and Wang-Fo marvelled at the length of those thin fingers, and hunted among his memories to discover whether he had not at some time painted a mediocre portrait or one of his ancestors that would now merit a sentence of death. But it seemed unlikely because Wang-Fo had not been an assiduous visitor at the Imperial Court. He preferred the farmers' huts or, in the cities, the courtesans' quarters and the taverns and the harbour where the dockers like to quarrel.

"You ask me what it is you have done, old Wang-Fo?" repeated the Emperor, inclining his slender neck toward the old man waiting attentively. "I will tell you. But, as another man's poison cannot enter our veins except through our nine openings, in order to show you your offences I must take you with me down the corridors of my memory and tell you the story of my life. My father had assembled a collection of your work and hidden it in the most secret chamber in the palace, because he judged that the people in your paintings should be concealed from the world since they cannot lower their eyes in the presence of profane viewers. It was in those same rooms that I was brought up, old Wang-Fo, surrounded by solitude. To prevent my innocence from being sullied by other human souls, the restless crowd of my future subjects had been driven away from me, and no one was allowed to pass my threshold, for fear that his or her shadow would stretch out and touch me. The few aged servants that were placed in my service showed themselves as little as possible; the hours turned in circles; the colours of your paintings bloomed in the first hours of the morning and grew pale at dusk. At night, when I was unable to sleep, I gazed at them, and for nearly ten years I gazed at them every night. During the day, sitting on a carpet whose design I knew by heart, I dreamed of the joys the future had in store for me. I imagined the world, with the Kingdom of Han at the centre, to be like the flat palm of my hand crossed by the fatal lines of the Five Rivers. Around it lay the sea in which monsters are born, and farther away the mountains that hold up the heavens. And to help me visualize these things I used your paintings. You made me believe that the sea looked like the vast sheet of water spread across your scrolls, so blue that if a stone were to fall into it, it would become a sapphire; that women opened and closed like flowers, like the creatures that come forward, pushed by the wind, along the paths of your painted gardens; and that the young, slim-waisted warriors who mount guard in the fortresses along the frontier were themselves like arrows that could pierce my heart. At sixteen I saw the doors that separated me from the world open once again; I climbed onto the balcony of my palace to look at the clouds, but they were far less beautiful than those in your sunsets. I ordered my litter; bounced along roads on which I had not foreseen either mud or stones, I travelled across the provinces of the Empire without ever finding your gardens full of women like fireflies, or a woman whose body was in itself a garden. The pebbles on the beach spoiled my taste for oceans; the blood of the tortured is less red than the pomegranate in your paintings; the village vermin prevented me from seeing the beauty of the rice fields; the flesh of mortal women disgusted me like the dead meat hanging from the butcher's hook, and the coarse laughter of my soldiers made me sick. You lied, Wang-Fo, you old impostor. The world is nothing but a mass of muddled colours thrown void by an insane painter, and smudged by our tears. The Kingdom of Han is not the most beautiful of kingdoms, and I am not the Emperor. The only empire which is worth reigning over is that which you alone can enter, old Wang, by the road of One Thousand Curves and Ten Thousand Colours. You alone reign peacefully over mountains covered in snow that cannot melt, and over fields of daffodils that cannot die. And that is why, Wang-Fo, I have conceived a punishment for you, for you whose enchantment has filled me with disgust at everything I own, and with desire for everything I shall possess. And in order to lock you up in the only cell from which there is no escape, I have decided to have your eyes burned out, because your eyes, Wang-Fo, are the two magic gates that open onto your kingdom. And as your hands are the two roads of ten forking paths that lead to the heart of your kingdom, I have decided to have your hands cut off. Have you understood, old Wang-Fo?"

Hearing the sentence, Ling, the disciple, tore from his belt an old knife and leaped toward the Emperor. Two guards immediately seized him. The Son of Heaven smiled and added, with a sigh: "And I also hate you, old Wang-Fo, because you have known how to make yourself beloved. Kill that dog."

Ling jumped to one side so that his blood would not stain his master's robe. One of the soldiers lifted his sword and Ling's head fell from his neck like a cut flower. The servants carried away the remains, and Wang-Fo, in despair, admired the beautiful scarlet stain that his disciple's blood made on the green stone floor.

The Emperor made a sign and two eunuchs wiped Wang-Fo's eyes.

"Listen, old Wang-Fo," said the Emperor, "and dry your tears, because this is not the time to weep. Your eyes must be clear so that the little light that is left to them is not clouded by your weeping. Because it is not only the grudge I bear you that makes me desire your death; it is not only the cruelty in my heart that makes me want to see you suffer. I have other plans, old Wang-Fo. I possess among your works a remarkable painting in which the mountains, the river estuary, and the sea reflect each other, on a very small scale certainly, but with a clarity that surpasses the real landscapes themselves, like objects reflected on the walls of a metal sphere. But that painting is unfinished, Wang-Fo; your masterpiece is but a sketch. No doubt, when you began your work, sitting in a solitary valley, you noticed a passing bird, or a child running after the bird. And the bird's beak or the child's cheeks made you forget the blue eyelids of the sea. You never finished the frills of the water's cloak, or the seaweed hair of the rocks. Wang-Fo, I want you to use the few hours of light that are left to you to finish this painting, which will thus contain the final secrets amassed during your long life. I know that your hands, about to fall, will not tremble on the silken cloth, and infinity will enter your work through those unhappy cuts. I know that your eyes, about to be put out, will discover bearings far beyond all human senses. This is my plan, old Wang-Fo, and I can force you to fulfill it. If you refuse, before blinding you, I will have all your paintings burned, and you will be like a father whose children are slaughtered and all hopes of posterity extinguished. However, believe, if you wish, that this last order stems from nothing but my kindness, because I know that the silken scroll is the only mistress you ever deigned to touch. And to offer you brushes, paints, and inks to occupy your last hours is like offering the favours of a harlot to a man condemned to death."

Upon a signal from the Emperor's little finger, two eunuchs respectfully brought forward the unfinished scroll on which Wang-Fo had outlined the image of the sea and the sky. Wang-Fo dried his tears and smiled, because that small sketch reminded him of his youth. Everything in it spoke of a fresh new spirit which Wang-Fo could no longer claim as his, and yet something was missing from it, because when Wang had painted it he had not yet looked long enough at the mountains or at the rocks bathing their naked flanks in the sea, and he had not yet penetrated deep enough into the sadness of the evening twilight. Wang-Fo selected one of the brushes which a slave held ready for him and began spreading wide strokes of blue onto the unfinished sea. A eunuch crouched by his feet, mixing the colours; he carried out his task with little skill, and more than ever Wang-Fo lamented the loss of his disciple Ling.

Wang began by adding a touch of pink to the tip of the wing of a cloud perched on a mountain. Then he painted onto the surface of the sea a few small lines that deepened the perfect feeling of calm. The jade floor became increasingly damp, but Wang-Fo, absorbed as he was in his painting, did not seem to notice that he was working with his feet in water.

The fragile rowboat grew under the strokes of the painter's brush and now occupied the entire foreground of the silken scroll. The rhythmic sound of the oars rose suddenly in the distance, quick and eager like the beating of wings. The sound came nearer, gently filling the whole room, then ceased, and a few trembling drops appeared on the boatman's oars. The red hot iron intended for Wang-Fo's eyes lay extinguished on the executioner's coals. The courtiers, motionless as etiquette required, stood in water up to their shoulders, trying to lift themselves onto the tips of their toes. The water finally reached the level of the imperial heart. The silence was so deep one could have heard a tear drop.

It was Ling. He wore his everyday robe, and his right sleeve still had a hole that he had not had time to mend that morning before the soldiers' arrival. But around his neck was tied a strange red scarf.

Wang-Fo said to him softly, while he continued painting, "I thought you were dead."

"You being alive," said Ling respectfully, "how could I have died?"

And he helped his master into the boat. The jade ceiling reflected itself in the water, so that Ling seemed to be inside a cave. The pigtails of submerged courtiers rippled up toward the surface like snakes, and the pale head of the Emperor floated like a lotus.

"Look at them," said Wang-Fo sadly. "These wretches will

die, if they are not dead already. I never thought there was enough water in the sea to drown an Emperor. What are we to do?"

"Master, have no fear," murmured the disciple. "They will soon be dry again and will not even remember that their sleeves were ever wet. Only the Emperor will keep in his heart a little of the bitterness of the sea. These people are not the kind to lose themselves inside a painting."

And he added: "The sea is calm, the wind high, the seabirds fly to their nests. Let us leave, master, and sail to the land beyond the waves."

"Let us leave," said the old painter.

Wang-Fo took hold of the helm, and Ling bent over the oars. The sound of rowing filled the room again, strong and steady like the beating of a heart. The level of the water dropped unnoticed around the large vertical rocks that became columns once more. Soon only a few puddles glistened in the hollows of the jade floor. The courtiers' robes were dry, but a few wisps of foam still clung to the hem of the Emperor's cloak.

The painting finished by Wang-Fo was leaning against a tapestry. A rowboat occupied the entire foreground. It drifted away little by little, leaving behind it a thin wake that smoothed out into the quiet sea. One could no longer make out the faces of the two men sitting in the boat, but one could still see Ling's red scarf and Wang-Fo's beard waving in the breeze.

The beating of the oars grew fainter, then ceased, blotted out by the distance. The Emperor, leaning forward, a hand over his eyes, watched Wang-Fo's boat sail away till it was nothing but an imperceptible dot in the paleness of the twilight. A golden mist rose and spread over the water. Finally the boat veered around a rock that stood at the gateway to the ocean; the shadow of a cliff fell across it; its wake disappeared from the deserted surface, and the painter Wang-Fo and his disciple Ling vanished forever on the jade-blue sea that Wang-Fo had just created.

Introduction ©1992 Alberto Manguel.

"What Has Literature Got To Do With It?" by Chinua Achebe. From *Hopes and Impediments* by Chinua Achebe. Copyright ©1988 by Chinua Achebe. Used by permission of Doubleday, a division of Bantam Doubleday Dell Publishing Group, Inc. and Harold Ober Associates.

"Two Words" by Isabel Allende. From *Black Water II: More Stories of the Fantastic*, Alberto Manguel (ed.). Originally published by Lester & Orpen Dennys Ltd. Copyright © by Isabel Allende. Translated from the Spanish by Alberto Manguel.

"The Boys Own Annual" from *Murder in the Dark*, Coach House Press, 1983. Copyright © Margaret Atwood.

"The Homecoming Stranger" by Bei Dao. From *Waves* by Bei Dao, William Heinemann Limited, 1987. Reprinted by permission of William Heinemann Limited.

"Everything and Nothing" by Jorge Luis Borges. From *Dreamtigers* by Jorge Luis Borges, translated by Mildred Boyer and Harold Morland. Copyright © 1964. Reprinted by permission of the University of Texas Press.

"A Commencement Address" from *Less Than One* by Joseph Brodsky. Copyright © 1986 by Joseph Brodsky. Reprinted by permission of Farrar, Straus & Giroux, Inc.

"A Letter to Our Son," by Peter Carey was first published in *Granta* 24, Summer 1988, and is reprinted by permission of the author and Rogers, Coleridge & White.

"A Woman Who Plays Trumpet Is Deported" by Michelle Cliff is from *Bodies of Water* by Michelle Cliff. This was reprinted with the permission of Michelle Cliff.

"Surface Textures" by Anita Desai. From *Winter Tales*, no. 23, Macmillan. Copyright © Anita Desai, 1978. Reprinted by permission of Rogers, Coleridge & White Ltd.

"Living Like Weasels" by Annie Dillard. From *Teaching a Stone to Talk*, by Annie Dillard. Copyright © 1982 by Annie Dillard. Reprinted by permission of HarperCollins Publishers.

"A Wedge of Shade" by Louise Erdrich. From *Louder than Words*, by William Shore, Editor. Copyright © 1989 by Louise Erdrich. Reprinted by permission of Vintage Books, a Division of Random House, Inc.

"Inside Memory" by Timothy Findley. From *Inside Memory: Pages From A Writer's Workbook* by Timothy Findley. Published by HarperCollins Publishers Limited, Toronto. ©1990 by Pebble Productions.

"Gotcha!" by Robert Fulford. From *Soho Square III* (London, 1990). ©1990 by Robert Fulford.

"From the Fifteenth District" by Mavis Gallant is from *From the Fifteenth District* by Mavis Gallant. Copyright © 1979 by Mavis Gallant. Reprinted by permission of the author and MacMillan Canada.

"A Writer's Freedom" by Nadine Gordimer. Reprinted by permission of Russell & Volkening, Inc. as agents for Nadine Gordimer. Copyright ©1975 by Nadine Gordimer.

"Summer Meditations" by Vaclav Havel is from *Summer Meditations* by Vaclav Havel, Alfred A. Knopf (Random House), 1992.

"The Prisoner Who Wore Glasses" by Bessie Head. Copyright © Bessie Head.

"The Second Coming of Come-by-Chance" by Janette Turner Hospital. From *Isobars*, by Janette Turner Hospital, published by McClelland and Stewart Inc., 1991. Reprinted by permission of the Canadian Publishers, McClelland & Stewart, Toronto, and Louisiana State University Press. [The poem "Come-by Chance" which appears in the story is by A. B. Paterson.]

"A Family Supper" by Kazuo Ishiguro first appeared in *Firebrand 2*. Copyright ©1982 Kazuo Ishiguro. Reprinted by permission of the author and Rogers, Coleridge & White.